EPC made Ezee

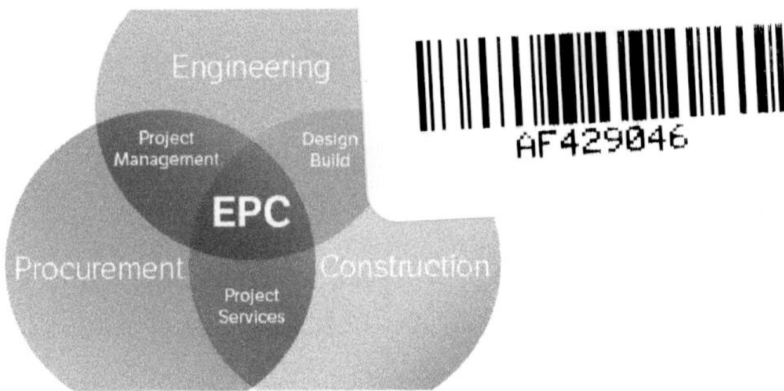

A Handbook to **EPC Execution** in the Energy Industry

SATHYAMURTHY VAMAN

ISBN 979-8-89415-628-6

Table of Contents

Preface

In the early days of my career, the world was a vastly different place. We relied on landline phones for communication, and the pursuit of knowledge was a physical journey. Accessing information meant visiting the Connemara library in Chennai (India) or browsing through books from street vendors along Mount Road. Moore Market, nestled near the bustling Madras (now Chennai) Central railway station, (India) was our treasure trove of affordable books. (Mostly used books)

These were simpler times, marked by the absence of the internet, email, and social media. The very concept of smartphones, Google, Amazon, Swiggy, and the remarkable capabilities of AI-driven tools like ChatGPT was beyond our wildest imagination.

Learning was a laborious process. To grasp new terminologies and technologies, we invested days and months of painstaking effort. Soft skill training, in particular, posed a formidable challenge. Mentorship was scarce, with knowledge often closely guarded, especially by those in positions of authority.

In these moments of retrospection, a profound saying from the Tamil language (The oldest Indian classical language) comes to my mind: "Naan petra inbum, peruga evv vaiyagham," attributed to the revered Saint Thirumoolar. It encapsulates a beautiful sentiment: "I share what I know, so the world may enjoy in bliss."

Inspired by this age-old wisdom, I found myself contemplating the idea of sharing the knowledge I've accumulated throughout my journey. As I reflect on the vast transformations that have reshaped our world, I am compelled to pass on what I've learned to the next generation.

This handbook embodies that intention and purpose. It serves as a bridge across generations, connecting the analog era of landlines and libraries to the digital frontier we now navigate. It stands as a testament to the evolution of information access and learning, with a deep commitment to the spirit of "Naan petra inbum, peruga evv vaiyagham."

Through its pages, we embark on a journey to share knowledge, embracing the digital revolution and striving to ensure that wisdom endures.

As we turn these pages, may we find inspiration in the past, enlightenment in the present, and a path forward into a future where knowledge knows no bounds?

About the Book

This is neither a textbook nor an academic book. It is the essence of three decades of hard work.

Audience

Who needs this handbook?

In the world of projects, whether you're crafting skyscrapers or baking cookies, this handbook is your trusty companion. It's like the Swiss Army knife of project management - versatile, indispensable, and surprisingly full of unexpected delights.

Who's the star of this show? Well, anyone who's remotely connected to any kind of project can tap into its wisdom.

Industry? Irrelevant.

Background? Doesn't matter.

You could be making software, building bridges, or teaching ducks to dance – this handbook's got something for you.

Now, here's the best part – it's a handbook for all, but not for everyone.

If you've been in the EPC (Engineering, Procurement, and Construction) game long enough to know the acronyms like the back of your hand, you might think this is child's play. And you know what? You're not wrong! But, hey, even the pros could use a little chuckle and a fresh perspective now and then.

How to read this book?

Now, let's talk about structure. This handbook is organized like a buffet, and you're the discerning diner. You don't need to feast on every chapter if you're racing against the clock. Pick what tantalizes your taste buds, and dig in. Each chapter is a culinary delight, complete on its own.

But here's the twist – it's not a ready reckoner. You won't find step-by-step recipes for every project predicament. Instead, it's your compass, guiding you through the labyrinth of project management, pointing out the pitfalls, and giving you a nudge in the right direction.

So, go ahead, and dive into the world of projects with a smile, a dash of humor, and this trusty handbook. Whether you're a greenhorn in the EPC jungle or a seasoned trailblazer, there's a nugget of wisdom here for everyone. Just remember, even in the serious business of projects, a little laughter and a lot of curiosity can take you a long way.

Note:

In India, there are lakhs of engineers who pass out of the college, every year, but only 20% are employable.

CHAPTER 1

Introduction to Oil & Gas Sector

What is Crude oil? and its composition

Crude oil is a complex mixture of hydrocarbons with varying molecular structures and properties. The composition of crude oil can vary significantly depending on the geological formation and location from which it is extracted. However, a typical composition of crude oil consists of the following components:

1. Hydrocarbons: The majority of crude oil is made up of hydrocarbons, which are organic compounds consisting of hydrogen and carbon atoms. These hydrocarbons can be categorized into three main types:

   - Paraffins (or alkanes): Straight-chained or branched hydrocarbons with single bonds between carbon atoms. They are relatively stable and have lower boiling points.
   - Naphthenes (or cycloalkanes): Cyclical hydrocarbons with single bonds between carbon atoms. They have slightly higher boiling points compared to paraffins.
   - Aromatics: Highly stable cyclic hydrocarbons containing a ring of carbon atoms with alternating double bonds. Aromatics have higher boiling points and often contribute to the odor and color of crude oil.

2. Sulfur Compounds: Crude oil contains varying levels of sulfur compounds, including hydrogen sulfide (H_2S) and organic sulfur compounds such as mercaptans and thiols. High sulfur content can pose environmental and processing challenges, and crude oil with low sulfur content is often preferred.

3. Nitrogen Compounds: Crude oil can also contain nitrogen compounds, including amines and amides. Nitrogen compounds can contribute to corrosion, fouling, and increased emissions during refining processes.

4. Oxygen Compounds: In addition to hydrocarbons, crude oil may contain small amounts of oxygen-containing compounds such as alcohols, aldehydes, and carboxylic acids. These compounds can increase the corrosiveness of the crude oil.

5. Trace Elements: Crude oil may contain trace amounts of metals such as vanadium, nickel, and iron, which can have detrimental effects on refining equipment and catalysts.

It's important to note that the specific composition of crude oil can vary widely depending on the geographical location and the specific oil field. Different crude oil sources have their unique characteristics, with variations in hydrocarbon ratios, sulfur content, and other impurities. This variation influences the suitability of crude oil for different refining processes and determines its market value.

The Donkey pump...

A donkey pump, also known as a beam pump or a nodding donkey, is a type of reciprocating pump commonly used in the oil and gas industry for the extraction of fluids, primarily crude oil, from wells. It operates using a nodding motion to create suction and lift the fluid to the surface.

Here's how a donkey pump typically operates in the oil and gas industry:

1. Pumping Unit: A donkey pump consists of a pumping unit, which comprises a walking beam, a horsehead, and a crankshaft. The pumping unit is installed above the wellhead and is connected to the well's production tubing.
2. Prime Mover: A prime mover, usually an electric motor or a reciprocating engine, provides power to operate the pumping unit. The prime mover drives the crankshaft, which is connected to the walking beam.
3. Sucker Rods and Polished Rod: The walking beam is connected to a set of sucker rods, which extend down into the wellbore, reaching the pump at the bottom. The sucker rods transmit the up-and-down motion from the walking beam to the pump.
4. Downward Stroke: As the walking beam moves downward, it pulls the sucker rods, which in turn pulls the polished rod at the surface. This downward motion creates a partial vacuum in the pump.
5. Upward Stroke: As the walking beam moves upward, it pushes the sucker rods, which push the polished rod. This upward motion compresses the fluid inside the pump and forces it up to the surface through the production tubing.
6. Fluid Extraction: The reciprocating motion of the donkey pump creates a cycle of suction and compression, allowing the fluid, such as crude oil, to be lifted from the wellbore to the surface. The extracted fluid flows through the production tubing and is collected in storage tanks or through a piping system for further processing.

7. Control and Monitoring: The operation of the donkey pump is often controlled and monitored through a central control system. This system can regulate the pumping speed, track production rates, monitor pump performance, and detect any anomalies or issues that may arise during the pumping process.

Donkey pumps are commonly used in both onshore and offshore oil and gas operations, particularly in areas where the reservoir pressure is insufficient to naturally lift the fluids to the surface. They offer a reliable and cost-effective method for the extraction of crude oil from wells.

The Christmas tree

In the oil and gas industry, a Christmas tree, also known as a wellhead or production tree, refers to a set of valves, fittings, and equipment installed at the top of an oil or gas well. It is named a "Christmas tree" due to its resemblance to a decorated tree when assembled with the various valves and equipment.

The Christmas tree serves as the primary interface between the wellbore and the surface, allowing for the control and management of fluids during drilling, production, and well-servicing operations. Here are the key components and functions of a typical Christmas tree:

1. Wellhead Housing: The Christmas tree is installed on top of the wellhead housing, which is a large metal casing embedded in the ground to provide structural support and sealing for the well.

2. Master Valve: The master valve, located at the top of the tree, is a large gate valve used to isolate and control the flow of fluids from the well. It is capable of quickly shutting down the well in case of emergencies or during maintenance activities.

3. Choke Valve: The choke valve is an adjustable valve positioned below the master valve. It regulates the flow rate of oil or gas from the well by controlling the size of the opening, which helps manage well pressure and optimize production.

4. Production Tubing: The Christmas tree typically has a production tubing spool, which connects the production tubing (a conduit for well fluids) to the wellhead. The tubing allows the extracted oil or gas to flow to the surface.

5. Wellhead Connectors: The Christmas tree includes various connectors and fittings to facilitate the attachment of ancillary equipment, such as pressure gauges, flow meters, or safety devices. These connectors provide access points for monitoring and maintenance.

6. Control Systems: Christmas trees are often equipped with control systems, including hydraulic or pneumatic actuators, to remotely operate the valves and control the flow of fluids. These control systems can be manually operated or linked to automated well control software.

The Christmas tree is an essential component in the oil and gas production process as it allows for the safe and controlled extraction of hydrocarbons from the well. It provides a means for well control, and pressure regulation, and connects the well with surface equipment and pipelines for further processing and transportation.

Sour and non-sour (Sweet) service

Sour and non-sour service are terms used in the oil and gas industry to describe the corrosiveness and potential hydrogen sulfide (H2S) content of fluids in various applications. Here's a breakdown of these terms:

1. Sour Service: Sour service refers to environments where fluids, typically hydrocarbons, contain significant levels of hydrogen sulfide (H2S) gas. H2S is a corrosive and toxic gas that can pose safety risks and lead to accelerated corrosion in equipment, pipelines, and vessels. Sour service conditions require the use of special materials and corrosion-resistant alloys to mitigate the corrosion effects caused by H2S.

Areas that may involve sour service include oil and gas production from reservoirs with higher sulfur content, refining facilities dealing with sour crude oil, and natural gas processing plants. Compliance with specific standards and regulations, such as NACE MR0175/ISO 15156, is necessary to ensure the safe handling and operation of equipment and systems in sour service environments.

2. Non-Sour Service: Non-sour service, also referred to as sweet service, pertains to applications or environments where hydrocarbon fluids do not contain significant amounts of H2S. Non-sour service conditions are typically associated with lower sulfur content in the fluid, minimizing the risk of corrosion and the need for specialized corrosion-resistant materials.

Non-sour service environments are commonly found in oil and gas production from fields with low sulfur content, processing of sweet or low-sulfur crude oil, and natural gas facilities where H2S concentrations are negligible or within permissible limits.

It's important to note that the distinction between sour and non-sour service is essential for the proper selection of materials, coatings, and operational practices to ensure equipment integrity, asset longevity, and the safety of personnel working in such environments. Adhering to industry standards and regulations and implementing appropriate corrosion control measures are crucial in both sour and non-sour service applications in the oil and gas industry.

What is LNG? and its composition

Liquefied Natural Gas (LNG) is primarily composed of methane, which is the main component of natural gas. However, its composition can vary slightly depending on the source of the natural gas and the liquefaction process used.

Here's a typical composition of LNG:

1. Methane (CH4): Methane is the primary component of LNG and typically accounts for around 85-95% of the total volume. It is a highly flammable hydrocarbon and is the main energy source in natural gas.
2. Ethane (C2H6): Ethane is usually present in smaller quantities in LNG, ranging from about 5-10%. It is often considered a valuable component due to its use as a feedstock for petrochemical processes.
3. Propane (C3H8) and Butane (C4H10): These hydrocarbons are also present in smaller amounts, typically around 1-5% each. Propane and butane can be separated from the LNG and used for various purposes, including heating and as fuel for cooking.
4. Traces of other Hydrocarbons: In addition to the primary components mentioned above, LNG may contain smaller traces of other hydrocarbons, such as pentane and hexane. These compounds are usually present in minimal quantities.

It's important to note that LNG is produced by cooling natural gas to extremely low temperatures (-162°C or -260°F), causing it to condense into a liquid state. This process removes impurities, such as water and carbon dioxide, as well as other gas components present in natural gas. The resulting LNG is predominantly composed of methane, with minor amounts of ethane, propane, butane, and other trace hydrocarbons.

The specific composition of LNG can vary based on region and the specific natural gas reserves from which it is produced. These variations may affect its physical properties and its suitability for specific applications, such as transportation, power generation, or use as a feedstock in industrial processes.

Major countries that produce oil and gas

In modern times, crude oil extraction has taken place in various regions around the world. The exploration and extraction of crude oil intensified with the growth of the global oil industry in the late 19th and early 20th centuries. Here are some notable locations and timeframes for modern crude oil extraction:

1. Pennsylvania, USA (1859): The modern petroleum industry began with the first commercial oil well drilled by Edwin Drake in Titusville, Pennsylvania, in 1859. This marked the birth of the oil industry in the United States and initiated the rapid growth of oil production.
2. Baku, Azerbaijan (late 19th century): The Baku oil fields in Azerbaijan were significant in the late 19th century, contributing to the growth of the global oil industry. It became one of the largest oil-producing regions in the world at the time.
3. Spindletop, Texas, USA (1901): The discovery of the Spindletop oil field in Texas, USA, in 1901 heralded the beginning of the modern oil boom in the United States. This significant oil discovery led to an exponential increase in oil production, solidifying the United States as a major player in the global oil industry.
4. Middle East (1930s onward): The Middle East, particularly countries like Saudi Arabia, Kuwait, Iran, and Iraq, became major contributors

to global oil production in the 1930s. Significant oil reserves were discovered, leading to large-scale extraction operations that continue to play a pivotal role in the global oil market.

5. North Sea (1970s): The development of the North Sea oil fields, primarily in the UK and Norway, began in the 1970s and significantly expanded global oil production. The North Sea region has become one of the most important offshore oil and gas production areas.

6. Offshore Gulf of Mexico (20th century): The Gulf of Mexico, particularly offshore drilling operations, has been a major area for crude oil extraction. The region has seen significant investment and exploration, contributing to the overall global oil supply.

7. Other notable regions: Other significant areas for modern crude oil extraction include Russia, Canada (particularly the oil sands in Alberta), Venezuela, Nigeria, Brazil, and various offshore fields around the world.

These locations represent some of the key areas where modern crude oil extraction has occurred, but it's important to note that oil extraction and production now take place in numerous countries across the globe. (Including India)

SOME OF THE ENGINEERING MARVELS OF THE WORLD

I embark on a journey through time to uncover the extraordinary engineering marvels that have left indelible marks on history. These towering achievements are not mere structures; they are the embodiment of human ambition, tenacity, and the art of turning dreams into reality. Join me as we delve into the fascinating world of some of the world's most iconic constructions, complete with the captivating technical details that powered their creation.

The Great Wall of China:

Picture ancient China, where the empire's destiny was etched into stone - the Great Wall. Spanning over 13,000 miles, this colossal fortification is a triumph of visionary planning and audacious project management. Comprising stone, brick, and tamped earth, it stands at an average height of 25 feet and a width of 15-30 feet. Watchtowers rise like sentinels, spaced at regular intervals, enabling communication across vast distances. It's a story of relentless coordination, resourcefulness, and strategic construction, which transformed disparate elements into a unified shield against invaders.

The Taj Mahal:

Now, let's journey to 17th-century India, where Shah Jahan's grief gave birth to the Taj Mahal, a jewel of love carved in marble. This timeless masterpiece is a symphony of precision and harmony. Crafted from locally quarried white marble, it reaches a height of 240 feet and is adorned with intricate inlay work of semi-precious stones, such as lapis lazuli and jade. Skilled architects, craftsmen, and laborers meticulously ensured that every detail found its place in this architectural marvel. The main dome alone measures 115 feet in height, an awe-inspiring testament to structural engineering.

The Brihadeeswarar Temple (Tanjore Temple):

Venturing India, southward, we discover the Brihadeeswarar Temple, an ancient marvel of Dravidian architecture. Constructed during the 11th century, this temple pushes the boundaries of innovation. Crafted entirely from granite, the temple's vimana (tower) soars to a height of 216 feet. A colossal monolithic granite dome, weighing approximately 81 tons, crowns this masterpiece, defying gravity and showcasing the audacity of the builders. The orchestra of architects, sculptors, and engineers played their parts to perfection, crafting a symphony of stone with meticulous precision. The temple was constructed using a technique called interlocking stones, Where the stones were carved to fit together without the use of mortar.

The Panama Canal:

Fast forward to the early 20th century, and we're in Panama, where the dream of connecting two oceans became a reality. The Panama Canal is more than a waterway; it's a testament to human determination and ingenious engineering. Locks, like colossal gates, raise and lower vessels, conquering steep terrain and elevational challenges. The canal spans approximately 50 miles, featuring a system of three locks at both the Atlantic and Pacific entrances. The Culebra Cut, a narrow passage through the Continental Divide, posed a monumental excavation challenge. Leadership, precision, and the art of moving millions of cubic meters of earth and rock defined this colossal project.

The Burj Khalifa:

Finally, we find ourselves in the heart of Dubai, gazing up at the Burj Khalifa, the world's tallest skyscraper. This modern marvel pierces the sky at a dizzying 2,717 feet. Its creation is a symphony of cutting-edge technology, high-strength concrete, and architectural innovation. The Burj Khalifa boasts 163 floors and features the world's fastest double-deck elevators. The massive concrete core, with its unprecedented continuous pour of concrete, underpins its structural integrity. The tower's exterior cladding, a technological marvel in itself, comprises over 103,000 square meters of reflective glazing and aluminum panels.

In this epic journey through time and space, we've encountered the wonders of the Great Wall of China, the Taj Mahal, the Tanjore Temple, the Panama Canal, and the Burj Khalifa. These awe-inspiring triumphs are not just structures; they are tales of human spirit, creativity, and the audacity of visionaries who dared to dream big. Project management, with its meticulous planning and technical expertise, served as the conductor, orchestrating the harmonious construction of these colossal dreams. These remarkable feats remind us that with unwavering determination, cutting-edge technology, and precise engineering, humanity can defy limits and leave its mark on the annals of history.

TITBITS

Mindfulness

Mindfulness is a type of meditation where you focus on being intensely aware of what you are sensing and feeling in the moment, without interpretation or judgment. It involves focusing your awareness on the present moment, often rooting yourself in your body through sensations.

Here are some key aspects of mindfulness:

1. **Being Present**: Mindfulness is about being fully engaged with whatever we are doing at the moment — free from distraction or judgment.
2. **Non-Judgmental Awareness**: It involves observing your thoughts and feelings without getting caught up in them.
3. **Focus on Sensations**: Mindfulness often involves focusing on sensations to root yourself in your body in the here and now.
4. **Regular Practice**: Mindfulness requires regular practice to gently retrain the mind to settle into the present moment.

Practicing mindfulness can have numerous benefits, including improving cognitive ability, reducing stress, anxiety, and depression symptoms, increasing a sense of well-being, and helping with pain management. It can be practiced during formal meditation or during everyday activities, like cooking, cleaning, or walking, or even during reading this book.!

TITBITS

How to prepare for an interview (entry level position)

1. **Understand the Job Description**: Thoroughly read the job description to understand what the organization is looking for in a candidate. Make a list of the skills, knowledge, and professional and personal qualities required for the job and identify at least one example of how you demonstrate each requirement.

2. **Research the Organization**: Understand the organization's mission, values, culture, and recent developments. This information can often be found on the company's website, social media platforms, and recent press releases.

3. **Prepare for Common Interview Questions**: Some common interview questions include "Tell me about yourself," "Why do you want to work here?" or "Describe a situation where…." Practice your responses to these questions. Make sure your answers are concise, specific, and job-related.

4. **Prepare Your Own Questions**: Interviews are a two-way process. Prepare thoughtful questions to understand if the organization is the right fit for you. You could ask about the team, projects, and growth opportunities.

5. **Dress Professionally**: First impressions matter. Even if the company has a casual dress code, it is better to be overdressed than underdressed for an interview.

6. **Plan Your Journey**: If the interview is at a physical location, plan your journey to ensure you arrive 10-15 minutes early. If it is a virtual interview, check your tech setup in advance.

7. **Follow Up**: After the interview, send a thank-you email to express your appreciation for the opportunity. This can help you stand out from other candidates.

The key to a successful interview is *preparation*. Good luck!

CHAPTER 2

Understanding the Fundamentals of Projects

Let us embark on a comprehensive exploration of the concept of "PROJECT," a term that has its origins rooted in Latin as "Projectum."

A project encompasses an individual or collaborative endeavor meticulously designed to achieve a specific objective. It is crucial to acknowledge that every project has a distinct beginning and a definitive conclusion.

This handbook focuses specifically on EPC projects, in the oil and gas sector, which represent Engineering, Procurement, and Construction contracts. We shall delve into a thorough comprehension of the project initiation and culmination processes.

The Conception Stage:

In this initial phase, an entity undertakes a geological survey aimed at identifying valuable resources, such as oil, gas, or rare minerals, within a specific geographic location. Following this discovery, the entity proceeds to develop a resource extraction concept driven by economic considerations, including aspects such as profit generation and job creation. This phase is appropriately referred to as the "conceptual stage."

Pre-FEED and FEED:

After the conceptual study is completed, the entity engages a contractor to conduct preliminary FEED (Front End Engineering Design) and/or feasibility studies. The preliminary FEED study yields a tentative model and cost estimation. If the entity determines that the project is viable, it moves forward with a more detailed "DEEP FEED," which provides a cost estimate within a range of -30% to +30%.

The output of the FEED phase serves as the basis for tender documentation, enabling the entity to invite contractors to undertake the actual construction activities outlined in the FEED.

With this foundation laid, it is now opportune to examine the various categories of contracts that come into play in project management.

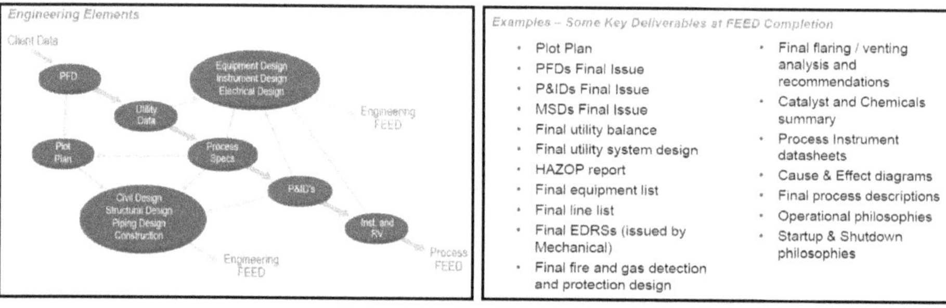

Project phases

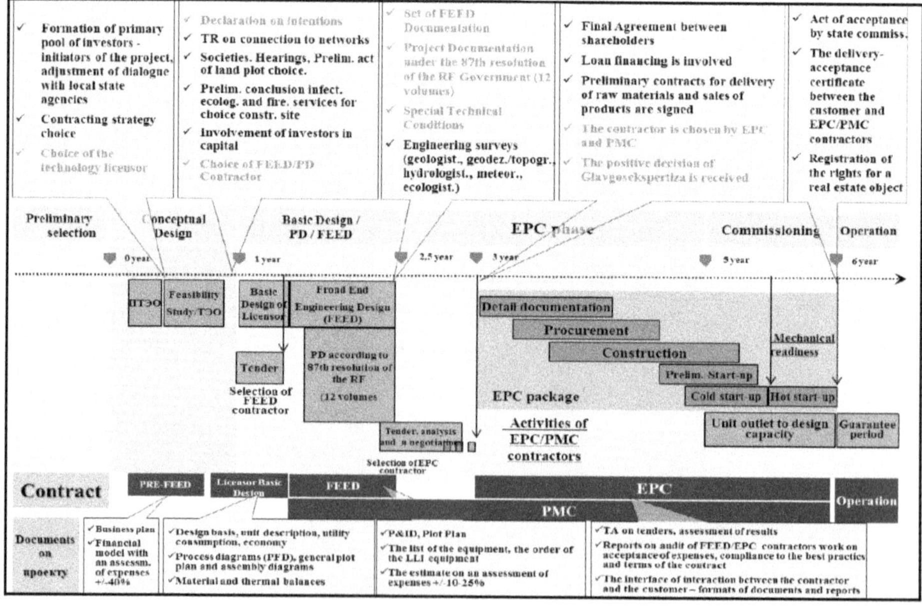

Estimating class

AACE Class	ANSI Classification	Typical Use	Project Definition	Expected Range of Accuracy		Other Terms
				Low Expected Actual Cost	High Expected Actual Cost	
Class 5	Order-of-Magnitude	Strategic Planning; Concept Screening	0% to 2%	-50% to -20%	+30% to +100%	ROM; Ballpark; Blue Sky; Ratio
Class 4		Feasibility Study	1% to 15%	-30% to -15%	+20% to +50%	Feasibility; Top-down; Screening; Pre-design
Class 3	Budgetary	Budgeting	10% to 40%	-20% to -10%	+10% to +30%	Budget; Basic Engineering Phase; Semi-detailed
Class 2	Definitive	Bidding; Project Controls; Change Management	30% to 75%	-15% to -5%	+5% to +20%	Engineering; Bid; Detailed Control; Forced Detail
Class 1		Bidding; Project Controls; Change Management	65% to 100%	-10% to -3%	+3% to +15%	Bottoms Up; Full Detail; Firm Price

Economic Analysis

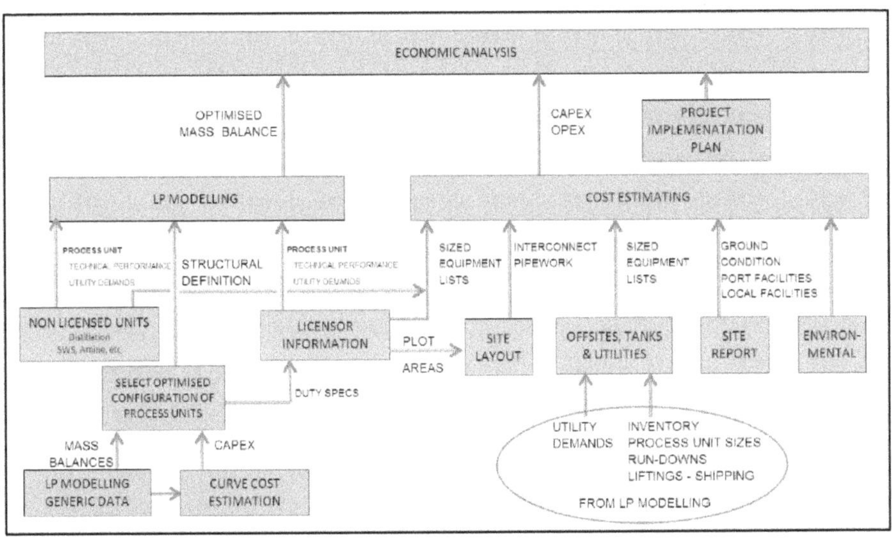

Final Investment Decision (FID)

A common term used about the last step of determining whether to move forward with the project by the Owner. (COMPANY)

THE EPC MODEL

Oil and gas, engineering, procurement, and construction (EPC) is a contract-based model that delivers a package of resources to complete infrastructure projects. The contractor carries out designing and detailed layout, onsite assembly, functional testing, procurement of equipment material, and manufacturing of systems. Oil and gas sectors rely on EPC contractors for long-term and large-scale projects that require skilled professional labor and fine-tuned project management. EPC contractors specialize in designing plans for aboveground storage tanks, power generation environmental controls, natural gas processing facilities, industrial power distribution, and material handling. It offers enhanced performance, flexibility, cost-effectiveness, and a single point of responsibility. EPC establishes a communication channel that allows the owners to manage all the relationships and infrastructure projects.

Oil & Gas EPC Market Trends:

The significant expansion in the oil and gas industry across the globe is one of the key factors driving the market growth. EPC is widely adopted for designing and project execution, such as building storage systems, drilling platforms, and advanced systems for exploration. In line with this, the rising power consumption, increasing population, and initiatives undertaken by the governments to generate electricity from renewable resources are favoring the market growth. Moreover, various technological advancements, such as the integration of the Internet of Things (IoT) for EPC contractors, are providing an impetus to the market growth. Additionally, the increasing demand for oil and gas EPC in the upstream sector as it offers fewer complexities, more accessibility to sites, lower investment requirements, and lower risk is positively impacting the market growth. Apart from this, the rapidly expanding automotive industry, the growing consumption of petroleum products, such as petrol, diesel, and CNG, and the implementation of various government initiatives to promote oil and gas projects are creating a positive outlook for the market.

> *The global Oil and Gas EPC market size was valued at USD 46.60 billion in 2021 and is projected to reach USD 92.81 billion by 2031. The global market is expected to grow at a Compound Annual Growth Rate (CAGR) of 7.2% over the forecast period 2022-2031.*

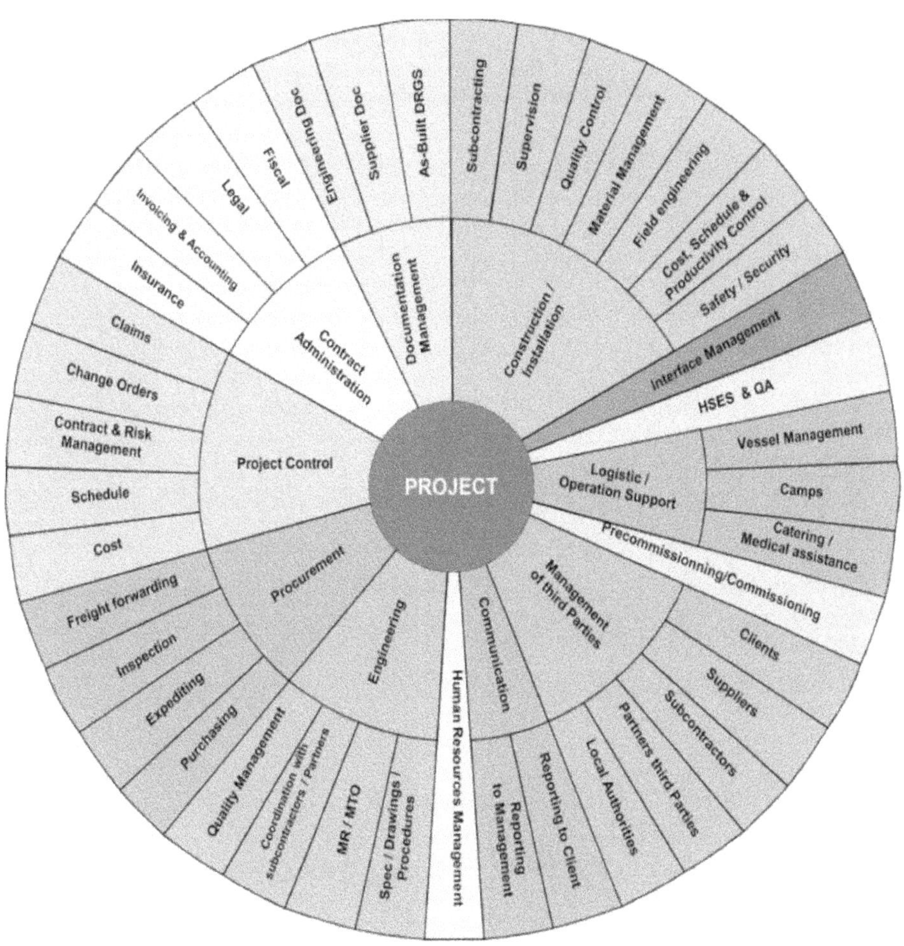

Exploring Various Types of EPC Contracts

In the realm of Engineering, Procurement, and Construction (EPC) contracts, a range of contractual arrangements exist to accommodate diverse project requirements. These contracts delineate the terms, responsibilities, and risk-sharing mechanisms between the project owner and the contractor. Below, we delve into the primary categories of EPC contracts:

Conventional Contracts:

Conventional EPC contracts encompass two fundamental types:

A. Fixed Price Contracts:

1. Lumpsum Contract (or Stipulated Price Contract, e.g., LSTK): Under this category, the contractor assumes full responsibility for executing the entire project at a predetermined price. They bear all associated risks and liabilities.
2. Unit Price Contract: Pricing in this model is established on a unit basis for specific project components.

B. Cost Reimbursable Contracts: This framework entails the project owner reimbursing the contractor for actual expenses incurred, along with a lump sum. The project owner assumes the project's risks. This category further bifurcates into:

1. Cost Plus Percentage of Cost Contract: This contract involves reimbursing costs incurred, along with an additional fee based on a percentage of the costs.
2. Cost Plus Fixed Fee Contract: The contractor is reimbursed for expenses incurred and receives a fixed fee for their services.
3. Cost Plus Incentive Fee: This arrangement entails reimbursement of costs incurred, along with an additional fee based on predetermined performance criteria.

4. Guaranteed Maximum Price: This contract sets a maximum limit on costs, with the contractor being responsible for any expenses exceeding the stipulated maximum.

Convertible Contracts:

In this distinctive category, known as Convertible Contracts, projects commence as cost-reimbursable arrangements and later transition into fixed-price contracts. This transition occurs at a later stage in the project's lifecycle, aligning with evolving project dynamics and risk assessments.

LNTP Contracts:

A limited notice to proceed (LNTP) is a notice issued by the Employer instructing the Contractor to proceed with a specific part of the work. This situation arises when not all conditions for fully proceeding with the project have been fulfilled.

BOOT Contracts:

BOOT contracts merge our expertise in design-build, own, operate, and transfer services with financing arrangements. It represents not just a service but a genuine long-term partnership.

BOT Contracts:

The BOT (build-operate-transfer) delivery model involves a private party that does not own the project as an asset. They receive a concession to operate it for a designated period.

While this overview provides a foundational understanding of EPC contract types, it is vital to explore each category in greater depth to fully grasp the nuances of their implementation. Regardless of the chosen contract type, the ultimate success of any project relies on skilled execution strategy and methodology. The guiding formula for success remains constant: executing a project within the defined budget, upholding optimal quality standards, and adhering to project timelines, resulting in a flourishing bottom line—a formula that leaves all stakeholders with a smile of satisfaction.

PROJECT TRIANGLE

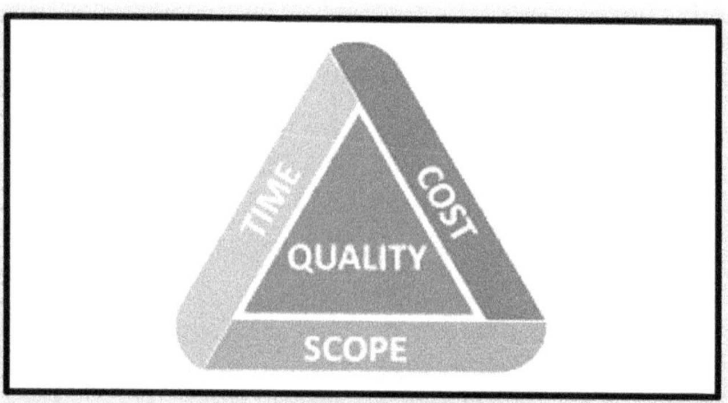

GREENFIELD AND BROWNFIELD PROJECTS:

Greenfield and brownfield projects are terms commonly used in the EPC (Engineering, Procurement, and Construction) industry to describe different types of projects based on their starting conditions and complexities.

Greenfield Projects:

Greenfield projects refer to projects that involve the development or construction of entirely new facilities or infrastructure on undeveloped or previously unused sites. These projects typically involve the construction of facilities from scratch, starting with an empty or "greenfield" site. They often require extensive planning, engineering design, procurement, and construction efforts.

Greenfield projects offer the opportunity to design and build facilities with full flexibility and customization to suit specific requirements. They involve the construction of new infrastructure, such as buildings, plants, pipelines, roads, and utilities, usually in areas where no previous installations exist. Greenfield projects may range from small-scale developments to large-scale industrial or commercial complexes.

Brownfield Projects:

Brownfield projects, on the other hand, involve the expansion, modification, or renovation of existing facilities or infrastructure. These projects take place on sites that already have some level of development or pre-existing installations. Brownfield projects often focus on the revitalization, upgrade, or repurposing of existing facilities to meet changing needs or comply with updated regulations.

Brownfield projects can involve various activities, such as the addition of new equipment, renovation of buildings, retrofitting of processes, or the integration of new technologies within an existing facility. They require careful planning, engineering assessments, and coordination to minimize disruptions to ongoing operations and ensure seamless transitions.

One of the primary challenges in brownfield projects is working within the constraints of the existing infrastructure, which may have space limitations, outdated systems, or conflicting configurations. However, brownfield projects also offer advantages such as potentially lower capital costs, reduced permitting requirements, and faster project timelines compared to greenfield projects.

Both greenfield and brownfield projects play significant roles in the EPC industry, catering to different development needs. Greenfield projects involve the construction of entirely new facilities, while brownfield projects focus on modifying and upgrading existing infrastructure. The choice between greenfield and brownfield depends on factors such as project objectives, available resources, site conditions, environmental considerations, and operational requirements.

THE EPC COST BREAKUP

In a typical EPC project, the cost break up is as follows (approximately, but may vary depending on the project scope and execution methodology)

Engineering – 5 % to 10%

Procurement – 40% to 50 %

Construction and commissioning – 30% to 40%

Even though the contribution of engineering is around 10% of the project cost, Engineering can <u>make or break</u> a project.

EPC execution earlier was like playing a test match in a cricket game, but now it is become a T20 match

THE WORLD'S LEADING OIL COMPANIES

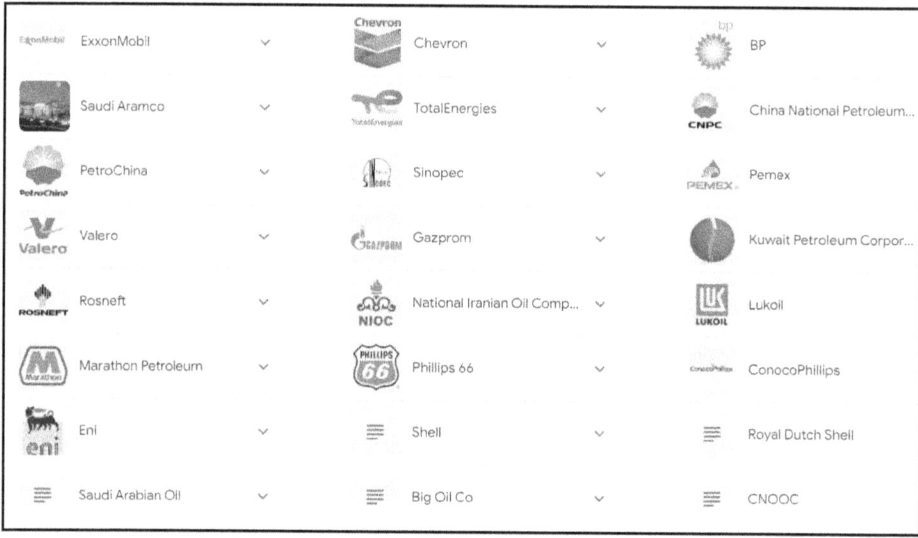

WORLD'S LEADING EPC CONTRACTORS

Petrofac	McDermott	Saipem
Samsung Engineering	Técnicas Reunidas	Bechtel
KBR	TechnipFMC	Fluor
JGC CORPORATION	NPCC	John Wood Group
Worley Limited	Hyundai Engineering	Technip
Hyundai Heavy Industries	Foster Wheeler	AtkinsRéalis
Saipem SpA	Galfar Engineering and C...	GS Engineering
Jacobs	Lamprell	Kent

OPPORTUNITIES FOR YOUNG ENGINEERS IN THE EPC INDUSTRY

Fresh engineers, whether diploma or degree holders, have entry-level opportunities in various roles within the oil and gas industry EPC. These entry-level positions allow them to gain valuable industry experience and contribute to the success of EPC projects. Some common entry-level opportunities for fresh engineers include:

1. Junior Engineer/Graduate Engineer: Many companies hire junior engineers or graduate engineers to support project teams in various disciplines such as mechanical, electrical, civil, or process engineering. These roles involve assisting senior engineers in design, analysis, documentation, and coordination tasks.

2. Field Engineer: Numerous EPC projects require field engineers to work on-site and oversee construction, installation, and commissioning activities. Field engineers collaborate with both contractors and clients to ensure project milestones are met and quality standards are upheld.

3. Project Coordinator/Assistant: Fresh engineers often start as project coordinators or project assistants, supporting project managers in coordinating activities, managing documentation, tracking project progress, and facilitating communication among different teams.

4. Estimator/Cost Engineer: Entry-level opportunities may also exist in roles related to cost estimation and control within EPC projects. Estimators or cost engineers assist in analyzing project requirements, developing cost estimates, evaluating bids, and monitoring project expenditures.

5. Quality Assurance/Quality Control Engineer: Fresh engineers can begin their careers in quality assurance or quality control roles, ensuring adherence to industry standards, conducting inspections, and implementing quality control processes.

6. Planning and Scheduling Engineer: Entry-level positions in planning and scheduling involve assisting in the development and maintenance of project schedules, monitoring progress, and analyzing potential delays or conflicts.

These entry-level opportunities provide a solid foundation for growth and advancement within the oil and gas industry EPC. Fresh engineers can acquire industry-specific knowledge, develop technical skills, and gain practical experience that will strengthen their career prospects in the long run. It is important to note that specific entry-level positions may vary depending on the company, project size, and location.

OPPORTUNITIES FOR YOUNG ENGINEERS TO UPGRADE IN THE EPC INDUSTRY

In the EPC (Engineering, Procurement, and Construction) industry within the oil and gas sector, there are several certifications available for individuals that demonstrate their knowledge, skills, and expertise in specific areas. Some of the different certifications available in the EPC industry for the oil and gas sector include:

1. Project Management Professional (PMP): The PMP certification, offered by the Project Management Institute (PMI), is widely recognized and focuses on project management skills, including planning, execution, monitoring, and control. It applies to individuals involved in managing EPC projects within the oil and gas sector.
2. Certified Cost Professional (CCP): The CCP certification, provided by the Association for Advancement of Cost Engineering (AACE International), validates the proficiency of professionals in cost engineering, estimating, and control within EPC projects in the oil and gas industry.
3. Certified Construction Manager (CCM): The CCM certification, offered by the Construction Management Association of America (CMAA), recognizes professionals who demonstrate competence in various aspects of construction management, encompassing planning, coordination, and control within EPC projects.
4. Certified Energy Manager (CEM): The CEM certification, granted by the Association of Energy Engineers (AEE), focuses on energy management and efficiency. It is relevant for individuals involved in EPC projects within the oil and gas industry with a focus on optimizing energy consumption and implementing sustainable practices.
5. American Welding Society (AWS) Certifications: The AWS offers various certifications related to welding, including Certified Welding Inspector (CWI) and Certified Welding Engineer (CWE). These certifications are beneficial for professionals involved in welding and fabrication within the EPC projects in the oil and gas sector.
6. Certified Safety Professional (CSP): The CSP certification, provided by the Board of Certified Safety Professionals (BCSP), validates individuals' knowledge and skills in managing occupational safety and health within EPC projects in the oil and gas industry.

7. API Certifications: The American Petroleum Institute (API) offers several certifications that are relevant to the oil and gas industry. These include certifications related to inspection, pressure vessel design, welding, quality management systems, and more.

These are just a few examples of certifications available in the EPC industry specifically within the oil and gas sector. It's important to note that specific roles and responsibilities within the industry may require different certifications depending on the individual's area of expertise, such as engineering disciplines, procurement, construction, quality management, etc. Additionally, industry-specific certifications from local professional bodies or organizations may also exist in different regions.

TITBITS

Walt Disney was told he lacked creativity.

One of the most creative geniuses of the 20th century was once fired from a newspaper because he was told he lacked creativity. Trying to persevere, Disney formed his first animation company, which was called Laugh-O-Gram Films. He raised $15,000 for the company but eventually was forced to close Laugh-O-Gram, following the close of an important distributor partner.

Desperate and out of money, Disney found his way to Hollywood and faced even more criticism and failure until finally, his first few classic films started to skyrocket in popularity.

CHAPTER 3

Project Management

As per Project Management Institute (PMI), project management is the application of knowledge, skills, tools, and techniques to meet project requirements

The primary components of a project management plan are:

1. Scope Statement
2. Critical Success Factors
3. Deliverables
4. Work Breakdown Structure
5. Schedule
6. Budget
7. Quality
8. Human Resources Plan
9. Stakeholder List
10. Communication
11. Risk Register
12. Procurement plan

In the context of project management, the fundamental components of a comprehensive project management plan include:

Scope Statement: A precise definition of the project's objectives and boundaries, serving as a foundation to prevent scope creep and ensure alignment with project goals.

Critical Success Factors: Key performance indicators that must be achieved for the project to be considered successful, guiding decision-making throughout the project's life cycle.

Deliverables: Tangible or intangible outcomes produced as a result of project activities, providing clear milestones to assess progress and goal attainment.

Work Breakdown Structure (WBS): A hierarchical breakdown of the project into manageable tasks or work packages, facilitating organization, assignment of responsibilities, and tracking.

Schedule: A structured timeline outlining the sequence and timing of project activities, preventing bottlenecks and delays by ensuring tasks are completed in a logical order.

Budget: A detailed financial plan that estimates costs for all project activities and resources, ensuring fiscal control and adherence to budget constraints.

Quality: Defined standards, criteria, and procedures to ensure project deliverables meet predetermined quality levels, emphasizing quality assurance and customer satisfaction.

Human Resources Plan: Identification of staffing requirements, roles, responsibilities, and skill sets necessary for effective project execution, ensuring the right personnel are allocated appropriately.

Stakeholder List: A comprehensive inventory of individuals and groups with an interest in or influence on the project, crucial for maintaining transparent communication and managing expectations.

Communication Plan: A structured approach detailing how project information will be disseminated, who will receive it, and communication frequency to keep stakeholders informed and engaged.

Risk Register: A documented catalog of potential project risks, including their likelihood and impact, accompanied by mitigation strategies to proactively address identified risks.

As mentioned before, this handbook places its focus solely on energy projects, particularly in the petroleum (oil and gas) domain. While there may be slight differences when executing other projects, the underlying fundamentals remain the same. It is important to note that we are not delving into the IT industry, where the concept of projects takes on an entirely different meaning. Our focus is specifically on industrial projects within the oil and gas industry.

Now, let's dive into understanding this industry in greater detail. In simple terms, the objective is to extract the oil and gas reserves beneath the Earth's surface and convert them into various products and by-products. However, in practice, this undertaking requires a multitude of technical skills, particularly when it comes to offshore operations conducted within the seas, which present additional challenges.

Let us now explore the distinctive categories that make up this vibrant sector.

Governance

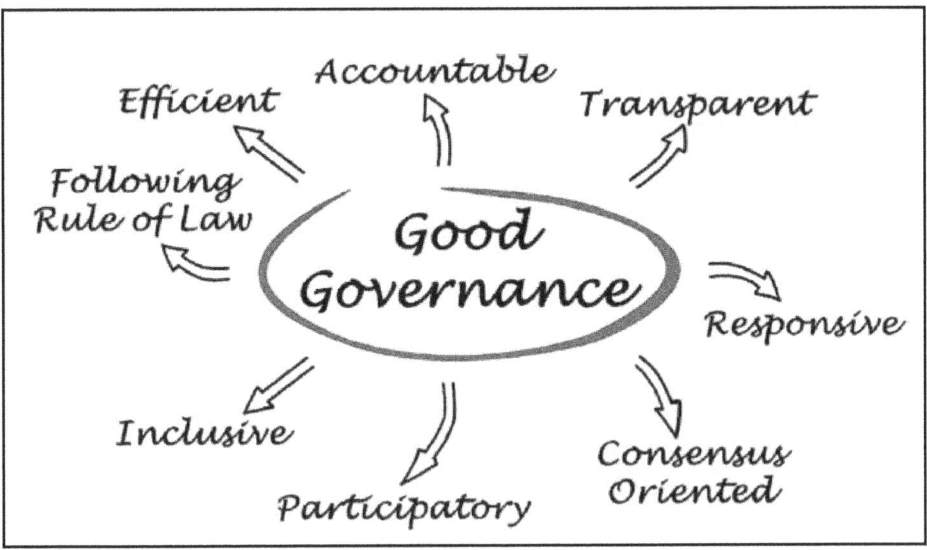

Personal attributes of a project manager

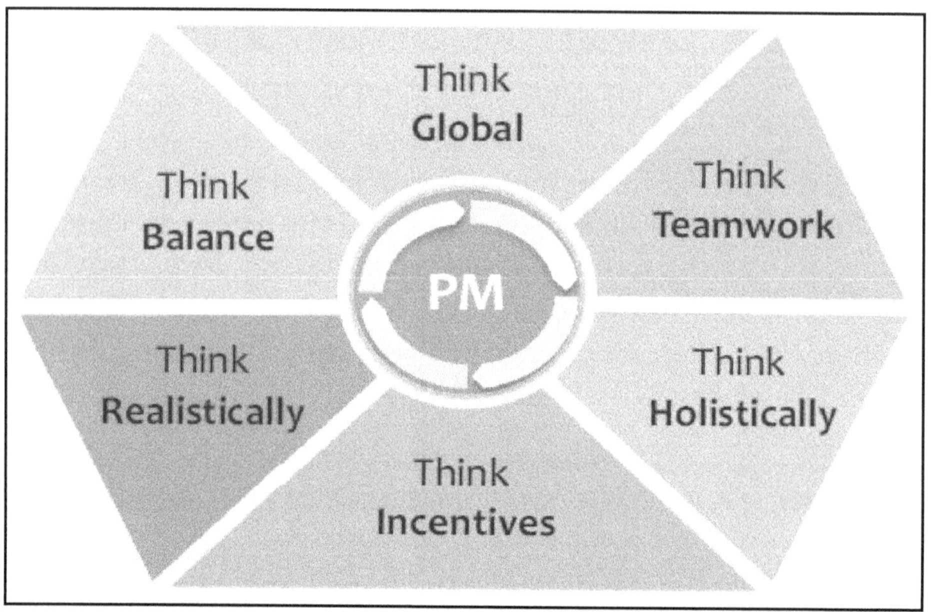

Project management areas

#	Knowledge Area	Description
1	Integration management	Enabling an integrated management of project management knowledge domains and processes across the life cycle
2	Stakeholders' management	Engaging people and groups who can be influential on the project or impacted by the project
3	Scope management	Delivering the project purpose and specific requirements
4	Schedule management	Managing time and sequence of tasks to ensure proper order of implementation
5	Resource management	Leading and directing people and other resources and assets required for project execution
6	Cost management	Planning and monitoring budgets, project expenses, cash flow, and other financial considerations

7	Communication management	Interacting with stakeholders to ensure a common understanding and minimize surprises
8	Risk management	Tackling uncertainties that may occur on projects, whether they are positive or negative events
9	Quality management	Examining the importance of quality including "fit for purpose"
10	Supply chain management	Managing the entire chain of resources for the project. The emphasis is on directing the acquisition, purchase, and allocation of external resources such as contractors, equipment, and other assets
11	Conflict management	Confronting escalating disagreements, which is a specific form of issue, can occur on intense projects with severe constraints such as time, schedule, and resources. Disagreements over task priorities, clashes among people, and different preferences for the chosen project management approach can emerge
12	Governance management	Establishing and implementing effective decision processes to prioritize, analyze, and make decisions can be vital for project success. Depending on the organizational complexity and the political environment, instituting the optimal governance processes and structures can greatly impact the successful execution of projects
13	Issue management (problems)	Addressing problems or obstacles quickly and effectively can minimize disruption to projects. Projects rarely, if ever, execute in an ideal state. Often, plans do not survive the first major challenge without causing change. Negative risks, such as threats, while they will ideally be successfully mitigated, can manifest and become issues

14	Adoption management	Facing organizational and people challenges early on provides more time and space for project managers to maneuver and create adoption plans. Studies have shown that most projects ultimately fail not because of technology but because of people. Project managers should proactively plan adoption upfront; this can reduce project inefficiency and possible failure
15	Operation management	Considering the project from a total life cycle perspective often includes the transitioning of project deliverables to operation. Project managers sometimes earn a poor reputation for focusing on the project implementation and neglecting what comes afterward. Most projects have significant post-project work, and the project team should be aware of this and plan for it accordingly in the transition planning

Project management phases

Project phase	Process
Ideation	Identify initial concerns and considerations about. (Knowledge domain)
Initiation	Identify and evaluate…. (Knowledge domain)
Preparation	Develop a plan for…. (Knowledge domain)
Implementation	Execute plan, monitor progress, and control of…. (Knowledge domain)
Transition and/or closure	Perform closure or transition activities for… (knowledge domain)

Business strategy framework

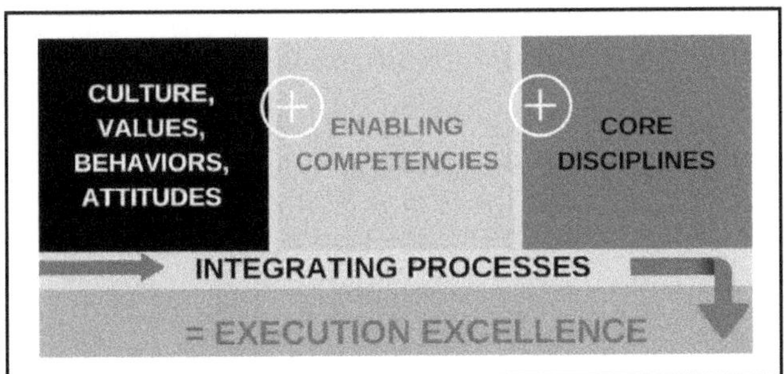

Roles and Responsibilities: (Typical to a large EPC project)

1. Project Sponsor:

- Provide overall strategic guidance and vision for the project.
- Secure necessary funding and resources.
- Ensure project objectives align with organizational goals.
- Provide support and assistance to the project team.

2. Project Director:

- Oversee and coordinate project activities.
- Develop and manage project plans, budgets, and schedules.
- Monitor project progress and ensure milestones are met.
- Provide leadership and direction to the project team.
- Communicate with stakeholders and address any issues or concerns.

3. Project Manager:

- Plan, execute, and close out the project.
- Develop project scope, goals, and deliverables.
- Coordinate with various departments and subcontractors.
- Monitor project timeline, budget, and resources.
- Identify and manage project risks and issues.
- Ensure project quality and client satisfaction.

4. Project Engineer:

- Provide technical expertise and guidance.
- Develop engineering specifications and drawings.
- Coordinate with design and procurement teams.
- Monitor and control engineering activities.
- Ensure compliance with industry standards and codes.
- Resolve technical issues and provide solutions.

5. Project Controls Manager:

- Develop and implement project control systems.
- Monitor project performance and progress.
- Manage project cost, schedule, and resource allocations.
- Develop and maintain project reports and dashboards.
- Analyze project data to identify trends and variances.
- Implement corrective measures as required.

6. Project IT Manager:

- Develop and manage project IT infrastructure.
- Coordinate with IT vendors and service providers.
- Implement and maintain project management software.
- Ensure data security and backup protocols are in place.
- Support project team with IT requirements.
- Troubleshoot and resolve IT issues.

7. Project Quality Assurance Manager:

- Develop and implement project quality management plans.
- Ensure compliance with quality standards and regulations.
- Conduct quality audits and inspections.
- Identify and resolve quality-related issues.
- Review and approve quality documentation.
- Develop and deliver quality training programs.

8. Project Quality Control Manager:

- Develop and implement quality control procedures.
- Monitor and inspect project activities for compliance.
- Conduct quality tests and inspections.
- Identify and address non-conformances.
- Collaborate with suppliers and subcontractors to ensure quality.
- Document and report quality control activities.

9. Project Engineering Manager:

- Oversee and coordinate engineering activities.
- Provide technical guidance and support to the engineering team.
- Ensure engineering deliverables are accurate and complete.
- Develop and maintain engineering standards and practices.
- Resolve technical issues and coordinate with other disciplines.
- Support project manager in resource planning and allocation.

10. Project Construction Manager:

- Plan and coordinate construction activities.
- Manage construction schedule, budget, and resources.
- Monitor site activities and ensure compliance with safety standards.
- Resolve construction-related issues and conflicts.
- Coordinate with subcontractors and suppliers.
- Ensure quality and timely completion of construction activities.

11. Project Pre-commissioning Manager:

- Develop and execute pre-commissioning plans and procedures.
- Coordinate with engineering and construction teams.
- Conduct pre-commissioning activities and tests.
- Ensure proper documentation and reporting.
- Coordinate with operation and maintenance teams for handover.
- Resolve any pre-commissioning issues or challenges.

12. Project Procurement Manager:

- Develop and implement project procurement strategy.
- Identify and select qualified suppliers and vendors.
- Negotiate and finalize procurement contracts.
- Monitor and expedite procurement activities.
- Ensure timely delivery of materials and equipment.
- Resolve procurement-related issues and disputes.

13. Project Engineering Lead Discipline Engineer:

- Lead and coordinate engineering activities for a specific discipline.
- Develop engineering specifications and standards.
- Provide technical guidance and support to the project team.
- Review and approve design documents and drawings.
- Coordinate with other discipline engineers.
- Ensure discipline-specific requirements and standards are met.

14. Project Interface Manager:

- Develop and implement project interface management plans and procedures.
- Identify key interfaces within the project and establish communication channels.
- Coordinate and facilitate communication and collaboration between different project stakeholders.
- Manage and resolve interface-related conflicts or issues.
- Ensure timely exchange of information and deliverables between different teams or departments.
- Coordinate and participate in interface meetings, workshops, and reviews.
- Monitor and track interface deliverables and milestones.
- Identify and mitigate interface risks or bottlenecks.
- Develop and maintain interface documentation and records.
- Provide interface-related support to the project team.

- Act as a liaison between the project and external parties, such as clients, subcontractors, or regulatory agencies.
- Ensure compliance with contractual obligations related to interfaces.
- Continuously improve interface management processes and lessons learned.

TITBITS

In any project, you'll encounter two interesting terminologies: DOCUMENT and RECORD. Let's dive into what they mean!

Documents, you see, are like the rulebooks. They eloquently describe how things should be done. They have their version history, going through revisions and updates. And oh boy, they require a proper review process before they can be considered officially updated.

Now, on the flip side, we have the enigmatic RECORDS. These babies show us how things were done. They serve as undeniable proof of the activities performed, a testament to whether those Standard Operating Procedures (SOPs) were diligently followed. And the best part? Records don't need any fancy release process. They speak for themselves!

According to ISO 9001, a DOCUMENT is vital information that must be maintained, while a RECORD is precious information that must be retained, forever capturing the essence of what took place.

But wait, there's more! Documents can be tweaked and revised, just like a fine recipe. They evolve as the project progresses. Records, on the other hand, are steadfast and unchangeable. They stand as irrefutable evidence of the past.

Let's highlight some examples to bring it all home. Documents can take the form of SOPs (Standard Operating Procedures), Manuals, Specifications, and all the meticulously crafted materials that guide us through the project. On the flip side, Records can be those trusty Checklists that we complete with gusto, those Reports that meticulously detail the project's progress, and other goodies that capture the essence of what went down.

In the EPC industry, we often refer to them collectively as "DELIVERABLES." That means all of the above, including those magnificent engineering drawings that bring these projects to life.

Joint Venture and Consortium

Joint ventures and consortiums are both forms of collaboration between companies or organizations for a specific project or business endeavor. However, there are distinct differences between these two terms:

Joint Venture:

A joint venture is a legal entity formed when two or more companies join forces to undertake a specific project or business activity together. In a joint venture, the participating entities contribute resources, expertise, and capital, and share risks, profits, and losses as agreed upon in a contractual agreement. The key characteristics of a joint venture include:

1. Shared Control and Ownership: All participating companies have joint control and ownership over the joint venture entity.
2. Mutually Agreed Objectives: The joint venture is created to fulfill specific mutually agreed objectives or goals, which can be short-term or long-term in nature.

3. Separate Legal Entity: A joint venture is often established as a separate legal entity, either as a partnership, limited liability company (LLC), or a corporation.
4. Sharing of Risks and Rewards: The participating companies share risks, costs, profits, and losses based on the agreed-upon ownership or shareholding percentages.

Consortium:

A consortium is a cooperative arrangement between multiple independent organizations or companies for a particular project or purpose. In a consortium, the participating entities collaborate and pool resources, expertise, and capabilities without forming a separate legal entity. The key characteristics of a consortium include:

1. Collaboration without Joint Ownership: Unlike a joint venture, a consortium does not involve joint ownership or control by the participating entities. Each organization operates independently, and shares resources, costs, and responsibilities based on the consortium agreement.
2. Common Objective or Project: The consortium is formed to achieve a specific project or objective, such as conducting research, developing technology, bidding for large contracts, or undertaking infrastructure projects.
3. Flexible Structure: A consortium is typically based on a contractual agreement that outlines the governance, roles, responsibilities, and resource-sharing arrangements among the participating entities. It allows for a flexible and adaptable collaboration structure.
4. Resource Pooling and Risk Sharing: Consortium members contribute resources, expertise, and funding towards the project or objective, and they share the associated risks, costs, and benefits.

In summary, the main difference between a joint venture and a consortium lies in the level of control, ownership, and legal structure.

A joint venture involves joint ownership and control, often with the establishment of a separate legal entity, while a consortium is a collaborative arrangement without joint ownership or a separate legal entity.

Good practices in project management:

1. Maintaining an action log:

   An action log shall be maintained by the project engineer, where all the major action points arising out of various meetings, and discussions are recorded. The review of this action log may be scheduled as a part of project progress meetings.

2. Minutes of meeting

   All meetings shall be recorded as minutes of meetings (MOM). This helps as a record in the future for various purposes, particularly during a conflict situation.

3. Key decision log:

   All key decisions by the project shall be recorded in the key decision log, which forms a record during the change management process.

4. Maintaining change log:

   All changes (commercial as well as technical) changes shall be recorded in the change log. This document is valuable for any claims in the future.

5. Document numbering log:

   This document is very important to avoid duplication of any numbering if document management system software is not used.

6. Lessons learned register:

   All lessons learned shall be collated in this register, and monitored for implementation, to avoid repeated errors.

7. Audit log:

   All audit results shall be collated in the audit log to ensure the implementation and effectiveness of the management systems.

8. Project diary:

   A project diary shall be maintained by each Lead, which helps in monitoring the work assignment to team members and the progress.

9. Summary of key project contractual obligations

   Even though a project execution plan includes the contractual requirements of a project, many project personnel do not read them thoroughly.

   Hence, a summary of the key contractual requirements/obligations can be prepared by the project manager and distributed to all stakeholders.

10. Project induction

    Project induction is an essential activity to be performed by the project management team once the project is kicked off.

    A typical induction may include the scope, schedule, quality and safety requirements, the technical specifications to be used, and the legal/regulatory requirements.

11. Team building sessions

    The project director must allocate a budget for team-building sessions at all phases of the project. This improves team spirit and collaboration.

12. Round table discussions

    With the advent of electronic systems, one-to-one or in-person interactions have come down drastically.

    If any activity requires multi-discipline inputs, a round table discussion can be called for and resolve the issues, across the table.

    Later this can be uploaded to the project document management system for record purposes.

13. Multi office execution

    Most of the projects have been executed using this model, in recent times. One project is executed from different offices across the globe.

    Standardization of the process(s) is the key in this case, and efforts need to be taken by the project management team, to ensure consistency of work execution methodology across all offices.

14. Stakeholder management

    It is the prime responsibility of the project management team to establish a cordial relationship with the stakeholders, for a smooth execution.

Project set up

A well-set project has a good chance of successful completion.

As soon as the project is awarded the project engineer and the quality assurance engineer must ensure the project is well set up.

A typical checklist for project setup is given below.

Within 60 days of the award, the QA personnel should conduct one project setup audit and list down the gaps.

Activities			**Completed**	**Not Required**
Transfer from Proposal				
A. Phase to Project Team				
A.1	Award Notification		☐	☐
A.2	LOI or Signed Contract		☐	☐
A.3	Contract development Meeting Notes		☐	☐
A.4	Specifications		☐	☐
A.5	Clarifications and Responses		☐	☐
A.6	Proposal Strategy		☐	☐
A.7	As-Sold Proposal Package		☐	☐
A.8	Estimate of Cost		☐	☐
A.9	Project Schedule/Milestones		☐	☐
A.10	Proposal PEP		☐	☐
A.11	As-Sold PROM		☐	☐
A.12	Subcontracts Bid Input		☐	☐
A.13	Procurement Bid Input		☐	☐

B. Project Administration			**Completed**	**Not Required**
B.1	Establish Key Org Structure		☐	☐
B.2	Define Roles and Responsibilities		☐	☐
B.3	Identify Team			
		Project Org Chart Issued	☐	☐
		Project Personnel Contact List issued	☐	☐
B.4	Mobilize Key Resources		☐	☐
B.5	Office Space			
B.6	Project Correspondence (Internal and External)			
		General correspondence set up complete	☐	☐
		Types and formats determined	☐	☐
		Templates provided	☐	☐
		Document distribution matrix developed	☐	☐
B.7	Organizing Project Documents			
		Determine project requirements	☐	☐
		Team access requirements	☐	☐
		Orientation help for project team	☐	☐
B.8	Establish File Index			
		Determine project requirements	☐	☐
		Team access requirements	☐	☐
		Orientation help for project team	☐	☐
B.9	Retention Schedule			
		Determine project requirements	☐	☐
		Communicate to project team	☐	☐
B.10	HR Policies		☐	☐
B.11	IT Systems & Set Up			
		Determine project requirements	☐	☐
		Team access requirements, e.g. network, SharePoint	☐	☐
		Orientation help for project team	☐	☐
B.12	Orientation & Training of PMT			
		Training needs assessed	☐	☐
		Project specific training plan developed	☐	☐

C. Project Alignment

C.1   Internal

- Review Scope and Execution Plan
- Overview of Project Baseline
- Project Profit Plan Initiatives
- Summary of Commercial Agreement
- Organization, Roles and Responsibilities
- Client Interface Management
- Lessons Learned
- Identify Issues and Action Plans

C.2   External (Client)

- Client: Overview and Schedule
- Client: Organization, Roles & Responsibilities
- Client: Risks and Value Drivers
- : Scope, Execution Plan & Schedule
- : Summary of Commercial Agreement
- : Organization, Roles and Responsibilities
- : Critical Success Factors
- Client & : Identify Issues and Action Plans

Project Scope Definition
D. (Include in PEP)

D.1   Description of Project

- Background
- History

D.2   Scope of Facilities

D.3   Scope of Services

D.4   Major Subcontract Scope

E. Project Controls

E.1   Coordinate/Develop Project Baseline

- Contract Summary
- Project Scope
- Project Execution Plan
- Level One Schedule
- Risk Assessment
- Project Estimate
- As Sold Pricing
- As Sold Budget
- Cash Flow
- Quality Indicators

E.2   Project Control Activities

- Level 0-3 Schedules prepared
- Project Staffing Plan prepared
- WBS & Standard Code of Accounts prepared
- 90 Day Schedule prepared
- Cost Control Baseline prepared
- Progress Control Baseline prepared
- Project Reviews and Reports - Set Up
  - Internal
  - External

F.	Project Risk Assessment		☐	☐
G.	Lessons Learned Review		☐	☐
H.	**Early Project Documentation**			
	H.1	Project Execution Overview	☐	☐
	H.2	Project Execution Plan (PEP) prepared	☐	☐
	H.3	Contract Summary prepared	☐	☐
	H.4	Project Profit Plan (PPP) prepared	☐	☐
	H.5	Project Quality Plan	☐	☐
	H.6	HSES Project Management Plan	☐	☐
	H.7	Project Procurement Plan	☐	☐
	H.8	Interface Management Plan	☐	☐

I.	**Discipline Readiness Checklists**			
	I.1	Project Controls	☐	☐
	I.2	Engineering	☐	☐
	I.3	Procurement	☐	☐
	I.4	Fab Operations	☐	☐
	I.5	Marine Operations	☐	☐
	I.6	Project Risk	☐	☐
	I.7	Quality	☐	☐
	I.8	HSES	☐	☐

TITBITS

Have you ever wondered about the distinction between CODE, STANDARD, and SPECIFICATION? These terms are all too familiar to engineers, and it's crucial to comprehend their nuances.

Let's start with CODE. Think of it as a set of widely accepted rules that dictate what needs to be done. It serves as a blueprint, guiding engineers in their work.

Moving on to STANDARDS, they provide the crucial "how to" aspect of implementing codes. Standards offer detailed instructions and best practices to ensure that codes are correctly executed.

Now, let's delve into SPECIFICATIONS. Unlike codes or standards, specifications outline the unique requirements of a particular company or product. They provide the specific criteria and characteristics that must be met.

MEP projects

MEP stands for Mechanical, Electrical, and Plumbing, and MEP projects refer to construction or infrastructure projects that involve the design, installation, and maintenance of mechanical, electrical, and plumbing systems within a building or facility. These systems are essential for the proper functioning and operation of any commercial, residential, or industrial structure.

1. Mechanical Systems:

- HVAC (Heating, Ventilation, and Air Conditioning) systems
- Fire suppression and protection systems
- Plumbing systems, such as water supply and drainage
- Elevators and escalators
- Building automation and control systems
- Lighting systems and fixtures
- Energy management systems

2. Electrical Systems:

- Power distribution systems
- Lighting systems (interior and exterior)
- Electrical panels and switchgear
- Fire alarm and detection systems
- Emergency lighting systems
- Uninterruptible power supply (UPS) systems
- Earthing and grounding systems

3. Plumbing Systems:

- Water supply systems (cold and hot water)
- Sanitary and wastewater drainage systems
- Fire sprinkler systems
- Stormwater management systems
- Plumbing fixtures, such as toilets, sinks, and showers

- Rainwater harvesting systems
- Irrigation systems (if applicable)

MEP projects involve the integration and coordination of these systems to ensure they operate efficiently, comply with regulatory standards, and meet the project requirements. MEP engineers and contractors work closely with architects, consultants, and other relevant stakeholders to design, install, and maintain these systems throughout the project lifecycle. The complexity and scale of MEP projects can vary widely, from small buildings to large-scale commercial complexes or industrial facilities.

TITBITS

Famous quotes by great people

The greatest glory in living lies not in never falling, but in rising every time we fall.

– Nelson Mandela

The way to get started is to quit talking and begin doing.

– Walt Disney

Your time is limited, so don't waste it living someone else's life. Don't be trapped by dogma – which is living with the results of other people's thinking.

– Steve Jobs

The future belongs to those who believe in the beauty of their dreams.

– Eleanor Roosevelt

You must be the change you wish to see in the world.

– Mahatma Gandhi

Spread love everywhere you go. Let no one ever come to you without leaving happier.

– Mother Teresa

The only thing we have to fear is fear itself.

– Franklin D. Roosevelt

Darkness cannot drive out darkness: only light can do that. Hate cannot drive out hate: only love can do that.

– Martin Luther King Jr.

In the end, it's not the years in your life that count. It's the life in your years.

– Abraham Lincoln

Science is a beautiful gift to humanity; we should not distort it.

– A. P. J. Abdul Kalam

CHAPTER 4

A Quick Glance at the Oil & Gas Industry

Onshore Oil and Gas Industry:

1. Upstream:

Exploration: The process of searching for oil and gas reserves beneath the Earth's surface.

Drilling: Creating wells to access underground reserves.

Production: Extracting and collecting oil and gas.

2. Midstream:

Transportation: Moving the extracted oil and gas from production sites to processing facilities.

Storage: Temporary holding of oil and gas before further processing.

Processing: Refining and cleaning the extracted resources.

3. Downstream:

Distribution: Transporting refined oil and gas products to end users.

Marketing: Selling and distributing petroleum products (e.g., gasoline, diesel) to consumers.

Retail: Selling refined products to the public through gas stations.

Classification of "ONSHORE" oil and gas industry

UPSTREAM	MIDSTREAM	DOWNSTREAM
Onshore exploration	onshore transportation	onshore distribution
onshore drilling	onshore storage	onshore marketing
onshore production	onshore processing	onshore retail

In the onshore industry, operations occur on land, from the exploration and drilling phases to the final distribution and retail of petroleum products. In contrast, the offshore industry involves extracting resources from beneath the seabed, with transportation to onshore facilities for processing.

Each segment - upstream, midstream, and downstream - plays a critical role in the overall oil and gas industry, ensuring that these valuable resources are efficiently and safely extracted, processed, and delivered to consumers worldwide.

Offshore Oil and Gas Industry:

1. Upstream:

Offshore Exploration: Searching for oil and gas reserves beneath the seabed.

Offshore Drilling: Creating underwater wells to access reserves.

Offshore Production: Extracting oil and gas from beneath the ocean floor.

2. Midstream:

Offshore Transportation: Transporting extracted resources from offshore platforms to onshore processing facilities.

Offshore Storage: Temporary storage of offshore-produced oil and gas.

3. Downstream:

Onshore Processing: Refining and cleaning the extracted offshore oil and gas.

Distribution: Transporting refined products to end users.

Marketing: Selling and distributing petroleum products.

Classification of "OFFSHORE" oil and gas industry

UPSTREAM	MIDSTREAM	DOWNSTREAM
Offshore exploration	Offshore transportation	Onshore processing
Offshore drilling	Offshore Storage	onshore distribution
Offshore production		onshore marketing
		onshore retail

In the offshore industry, space is a constraint, and hence the layout plays a major role. Also, Weight control is essential to optimize the cost

In the case of the onshore industry, such constraints do not exist to that extent, but safety and environmental requirements are high.

The material of construction differs in both cases, depending on the planned life of the installation.

SURF in the Oil and Gas Industry:

Subsea Umbilical, Risers, and Flowlines (SURF) represent a vital component of the modern oil and gas industry. As the demand for hydrocarbon resources continues to grow, extracting oil and gas from increasingly challenging offshore environments becomes imperative. SURF systems play a pivotal role in this endeavor, enabling safe, efficient, and reliable operations beneath the ocean's surface. Now, we will explore the significance of SURF in the oil and gas sector, its key components, and its contribution to the industry's overall success.

The Significance of SURF:

SURF is a collective term encompassing three essential elements:

Umbilicals: These are composite or hybrid cables that transport hydraulic fluids, electrical power, and data signals to and from the subsea equipment. Umbilicals serve as the lifeline that connects offshore installations with control systems on the surface.

Risers: Risers are conduits that transport oil and gas from the seabed to production facilities or vessels on the surface. They serve as critical arteries for hydrocarbon transportation in offshore drilling operations.

Flowlines: Flowlines are pipelines that carry oil and gas from the wellhead to processing facilities or export terminals. They are responsible for the efficient and safe transfer of hydrocarbons.

Key Components of SURF: SURF systems are complex and technologically advanced. Key components include:

Control Systems: These systems manage the operation of subsea equipment, controlling functions such as wellhead valves and subsea boosting pumps.

Umbilical Termination Assembly (UTA): The UTA connects the umbilical to subsea infrastructure, allowing for the transfer of fluids and signals.

Riser Systems: These systems consist of flexible or rigid risers, each tailored to specific environmental and operational conditions.

Flowline Infrastructure: Flowlines can vary in design, including flexible, rigid, or hybrid solutions to suit subsea challenges.

Contribution to the Industry's Success: SURF plays a pivotal role in the success of the oil and gas industry in several ways:

Maximizing Recovery: Offshore hydrocarbon reserves are often located in challenging environments, including deepwater and ultra-deepwater. SURF technology enables the extraction of these reserves, maximizing production and recovery rates.

Safety and Reliability: SURF systems are engineered for durability and reliability. They help minimize downtime, reduce operational risks, and enhance safety for personnel and the environment.

Environmental Responsibility: By enabling subsea tiebacks and reducing the need for surface infrastructure, SURF systems contribute to a smaller environmental footprint and more sustainable operations.

Efficiency and Cost-effectiveness: SURF technologies optimize offshore operations, improving efficiency and reducing operational costs over the life of a project.

In the dynamic and challenging realm of offshore oil and gas exploration and production, SURF systems serve as a critical enabler of success. These intricate and technologically advanced systems ensure safe, reliable, and efficient subsea operations, unlocking previously inaccessible hydrocarbon resources and contributing to the industry's sustainable growth. As offshore operations continue to evolve and expand, the role of SURF remains indispensable, representing a testament to human ingenuity and innovation in the pursuit of energy security and environmental stewardship.

FPSO

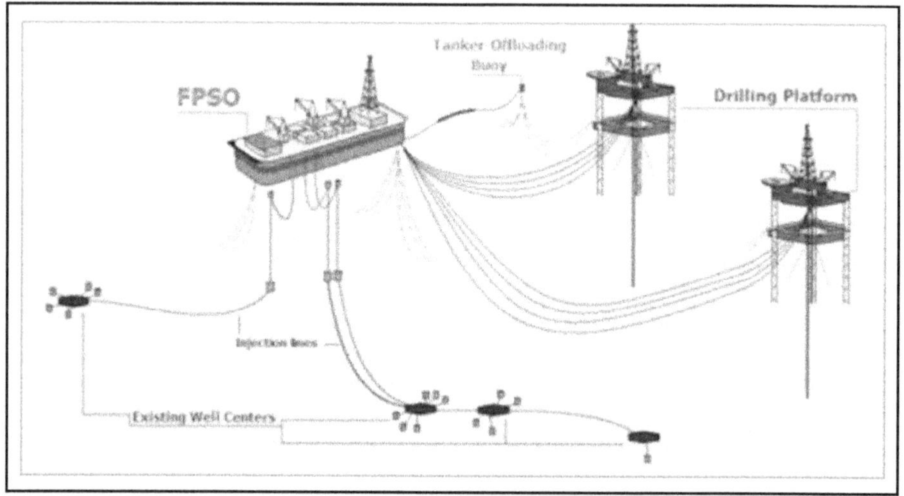

FPSO definition

As onshore oil discoveries continue to decline, FPSOs will become increasingly vital for the oil and gas industry. There are more than 200 FPSOs today operating around the globe. They're less expensive than traditional offshore oil and gas platforms, more flexible, safer, and time-efficient. Here is a breakdown of the FPSO acronym:

Production –

The "P" in FPSO is what separates these vessels from FSOs. Production refers to the processing of oil and gas. Hydrocarbons are produced in seabed wells, and this is transported to the FPSO via flowlines and risers. The hydrocarbons are then separated into oil, gas, and impurities via the production facilities on the deck of the FPSO.

Flowlines – Flowlines carry hydrocarbons directly from seabed wells. These can be flexible or rigid.

Risers – Developed for vertical transportation. This is the section of the line from the seabed to the topside.

Storage – Once the oil has been processed, it is transferred to cargo tanks in the double hull of the vessel.

Offloading – Offloading refers to transferring the gathered contents to additional transfer conduits. Crude oil that is stored in the vessel is then transferred to tankers and pipelines heading ashore. Gas is either transported to the shore via pipeline or recycled back into the field to increase production.

One of the typical features of an FPSO is the turret mooring system, which is usually fitted inside and integrated into the FPSO's hull. This is a Bluewater core technology. The turret is moored to the seabed with chains, wires, and anchors and has bearings allowing unrestricted 360° rotation of the FPSO around the turret ("weathervaning"). The FPSO will normally lay head to the prevailing environment.

The living quarters provide accommodation for offshore personnel and contain the temporary refuge, (emergency) control rooms, offices, and dining & recreation lounges.

FLNG

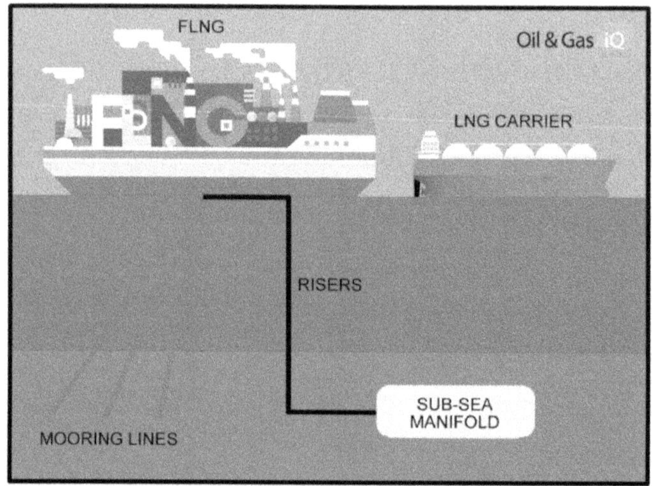

The FLNG facility is moored directly above the natural gas field. It routes gas from the field to the facility via risers. The gas is then processed and treated to remove impurities and liquefied through freezing, before being stored in the hull. Ocean-going carriers will offload the LNG, as well as the other liquid by-products, for delivery to markets worldwide. The conventional alternative to this would be to pump gas through pipelines to a shore-based facility for liquefaction, before transferring the gas for delivery.

Subsea Pipelines and Associated Components:

Subsea pipelines and associated components are the unsung heroes of the oil and gas industry, quietly facilitating the transportation of hydrocarbons from beneath the ocean floor to processing facilities and beyond. These intricate systems represent a critical link in the energy supply chain, enabling the safe, efficient, and reliable movement of oil and gas resources. We will now explore the significance of subsea pipelines, their key components, and their pivotal role in the global energy landscape.

The Significance of Subsea Pipelines:

Subsea pipelines are the arteries of offshore oil and gas operations, serving as the primary mode of transportation for hydrocarbons from subsea wells to onshore facilities or export terminals. Their significance lies in several key areas:

Efficient Transportation:

Subsea pipelines are designed to transport oil and gas over long distances efficiently. They eliminate the need for costly and environmentally impactful tanker transportation.

Safety and Environmental Responsibility:

By minimizing the exposure of hydrocarbons to the open environment, subsea pipelines enhance safety and reduce the risk of oil spills, aligning with industry commitments to environmental responsibility.

Maximizing Recovery:

These pipelines enable the extraction of offshore reserves, including those in challenging deepwater and ultra-deepwater locations, contributing to maximizing recovery rates.

Key Components of Subsea Pipelines:

Subsea pipelines are complex systems composed of various components, each with a specific function:

Pipelines:

The core component, pipelines, are typically made of high-strength materials such as steel. They transport oil and gas from the wellhead to processing facilities or export terminals.

Pigging Systems:

Pigs, robotic devices that move through the pipelines, are used for cleaning, inspection, and maintenance. They help ensure the integrity and efficiency of the pipelines.

Valves and Controls:

Subsea valves and control systems are strategically placed along the pipeline route. These components manage the flow, pressure, and integrity of the transported hydrocarbons.

Risers:

Riser systems connect subsea pipelines to surface facilities, allowing for the safe and efficient transfer of oil and gas. Flexible and rigid risers are utilized depending on environmental and operational conditions.

The Role of Subsea Pipelines in the Industry's Success: Subsea pipelines play a pivotal role in the oil and gas industry's success:

Infrastructure Backbone: They form the infrastructure backbone for offshore drilling operations, enabling the efficient evacuation of oil and gas from wellheads.

Environmental Stewardship: By reducing the need for surface infrastructure and minimizing the environmental footprint, subsea pipelines contribute to sustainable oil and gas production.

Cost Efficiency: They offer a cost-effective mode of transportation compared to alternative methods like tanker shipments, which are often more expensive and pose environmental risks.

Reliability: Subsea pipeline systems are engineered for durability and reliability, with regular inspection and maintenance ensuring long-term operational efficiency.

Subsea pipelines and associated components are the lifelines of the oil and gas industry, silently powering the global energy supply. Their efficient and environmentally responsible transportation of hydrocarbons, often from challenging offshore environments, underscores their significance. As the industry continues to evolve and address sustainability concerns, subsea pipelines remain a cornerstone of responsible energy production, enhancing safety, efficiency, and cost-effectiveness. In a world ever-hungry for energy, these submerged marvels continue to demonstrate their indispensable role in the global energy landscape.

Piping Vs Pipeline – the difference

Factor	Plant Piping	Pipeline Piping
Location	Within industrial facilities	Extends over long distances
Purpose	Internal fluid transport within a plant	Long-distance transmission of fluids
Applicable Codes	ASME B31.3 (Process Piping)	ASME B31.4 (Liquid Petroleum Transportation Piping) ASME B31.8 (Gas Transmission and Distribution Piping)
Material of Construction	Various materials depending on process and requirements	Steel is common (e.g., API 5L for oil and gas pipelines) Other materials like HDPE for certain applications
Pipe Size and Complexity	Varies based on process and facility requirements	Typically, larger diameter pipes for long-distance transmission
Design Pressure/ Operating Conditions	Usually moderate to high pressure within the plant	High pressure for long-distance transmission
Terrain Considerations	Enclosed within industrial facilities	May traverse varied terrains: land, underwater, underground
Regulatory Considerations	Plant-specific codes, environmental and safety regulations	Pipeline-specific codes, integrity management systems, environmental and safety regulations

It's important to note that the codes mentioned here are representative examples and may vary depending on the specific country, region, or industry standards. Additionally, material selection for both types of piping depends on factors such as fluid transport, temperature, pressure, corrosion considerations, and project requirements. Therefore, the actual material and codes used can vary based on project-specific engineering and design requirements

Engineering criticality assessment

Engineering Criticality Assessment: A Proactive Approach to Ensuring Safety and Reliability in Cross-Country Pipeline Installation

In the field of cross-country pipeline installation, safety and reliability are of utmost importance. The successful operation of these pipelines is crucial for the transportation of various substances, such as oil, gas, or water, over long distances. To ensure the integrity and functionality of these pipelines, engineers employ Engineering Criticality Assessment (ECA) as a proactive approach to identify potential risks and critical points that could lead to failures. We will explore the significance of ECA in the context of cross-country pipeline installation, its process, and its benefits in maximizing safety and reliability.

The Significance of Engineering Criticality Assessment in Cross-Country Pipeline Installation:

Cross-country pipelines represent a complex network of components and materials that necessitate rigorous evaluation to maintain safety and reliability. Engineering Criticality Assessment plays a vital role in this process by systematically identifying high-risk areas or components within the pipeline system. By conducting ECA, engineers can comprehensively understand the potential risks, likelihood of failures, and their ramifications. This enables them to prioritize corrective measures, allocate resources efficiently, and develop strategies to minimize or eliminate the identified critical issues. As a result, ECA becomes an essential component in ensuring the overall safety and reliability of cross-country pipeline installations.

The Process of Engineering Criticality Assessment for Cross-Country Pipelines:

The ECA process for cross-country pipeline installations involves several key steps. Firstly, engineers identify the critical components of the pipeline, such as weld joints, valves, or high-stress areas, considering factors such as operating conditions, material properties, and environmental impacts. This identification is typically performed through detailed design analysis and evaluation of potential failure modes. Subsequently, engineers assess the criticality of these components by analyzing their failure modes, estimating their probabilities, and evaluating potential consequences. Techniques like Failure Modes and Effects Analysis (FMEA) and Fault Tree Analysis (FTA) are commonly employed to analyze the criticality levels. Finally, engineers prioritize these critical areas and develop effective strategies, including regular inspection and maintenance programs, appropriate material selection, and monitoring systems, to mitigate the identified issues.

Benefits of Engineering Criticality Assessment in Cross-Country Pipeline Installation:

Implementing ECA brings numerous benefits to cross-country pipeline installations. Firstly, it enables engineers to gain a holistic understanding of potential risks and failures within the pipeline system, facilitating the identification of areas that require immediate attention. With this knowledge, resources can be allocated efficiently, directing efforts towards critical components, thereby minimizing the chances of catastrophic failures. Secondly, ECA allows engineers to develop targeted strategies for risk mitigation, which enhance the overall safety and reliability of the pipeline installation. By addressing critical points, the probability of failures is significantly reduced, ensuring seamless operations. Moreover, ECA assists engineers in complying with industry standards and regulations related to safety and reliability, averting legal issues, and maintaining the reputation of the pipeline installation. Lastly, conducting ECA fosters a proactive culture within engineering teams, making the identification and resolution of critical issues an integral part of pipeline design and development.

Engineering Criticality Assessment is a crucial aspect of ensuring safety and reliability in cross-country pipeline installations. By proactively evaluating and addressing potential risks and failures, engineers can minimize the occurrence of catastrophic incidents and enhance the performance of the pipelines. The ECA process involves identifying critical components, assessing their potential impact, and developing strategies for risk mitigation. The benefits of ECA include efficient resource allocation, targeted risk mitigation, compliance with regulations, and a proactive approach toward safety and reliability. As cross-country pipeline installations continue to expand, the significance of ECA will only grow, driven by the imperative to minimize failures and enhance the trustworthiness of pipeline systems.

Impact of Non-Destructive Testing on Engineering Criticality Assessment in Cross-Country Pipeline Installation

Non-destructive testing (NDT) techniques play a crucial role in ensuring the safety and reliability of cross-country pipeline installations. As a vital component of the Engineering Criticality Assessment (ECA) process, NDT helps engineers thoroughly evaluate the integrity of critical components within the pipeline system. We will explore the impact of NDT on ECA in the context of cross-country pipeline installation and highlight its significance in enhancing safety and reliability.

The Role of Non-Destructive Testing in Engineering Criticality Assessment:

Non-destructive testing techniques provide engineers with valuable insights into the condition and integrity of critical components within cross-country pipelines. NDT methods allow for the detection and evaluation of defects, anomalies, corrosion, and other potential issues that may compromise the performance and safety of the pipeline system. By incorporating NDT techniques into the ECA process, engineers can gain a comprehensive understanding of the criticality of various components and make informed decisions regarding risk mitigation strategies.

The Impact of Non-Destructive Testing on Engineering Criticality Assessment:

The utilization of non-destructive testing within the ECA process has several significant impacts on engineering criticality assessment for cross-country pipeline installations. Firstly, NDT techniques enable engineers to detect and locate defects and anomalies in critical components without causing any damage or disruption to the pipeline system. This non-intrusive approach allows for the identification of potential areas of concern, such as cracks or corrosion, which can adversely impact the integrity of the pipeline. By detecting these issues early on, engineers can take appropriate measures to address them, minimizing the likelihood of failure and improving overall safety.

Secondly, NDT methods provide engineers with quantitative data regarding the severity and size of defects within critical components. This information is invaluable for assessing the criticality of those components accurately. By understanding the size and severity of defects, engineers can determine the potential risks and consequences associated with failures at these locations. This data guides decision-making processes, enabling engineers to prioritize actions and allocate resources effectively to mitigate the risks.

Thirdly, non-destructive testing allows for continuous monitoring and evaluation of critical components throughout the lifecycle of the pipeline system. Regular inspections with NDT techniques provide engineers with up-to-date information on the condition of these components, helping them identify any degradation or changes over time. Timely detection of issues allows for proactive maintenance and ensures the ongoing safety and reliability of the pipeline system.

Additionally, the integration of NDT techniques with engineering criticality assessments provides valuable inputs for designing and implementing appropriate inspection and maintenance programs. By considering the information gathered through NDT, engineers can establish optimal inspection intervals, determine the most suitable inspection methods, and effectively plan maintenance activities. This approach increases the efficiency and accuracy of risk mitigation strategies and contributes to the long-term safety and reliability of the pipeline installation.

Non-destructive testing techniques have a significant impact on engineering criticality assessment in cross-country pipeline installations. Through the use of NDT methods, engineers can detect and evaluate defects, anomalies, and corrosion within critical components without causing damage to the pipeline system. The quantitative data provided by NDT assists in determining the severity and size of defects, enabling accurate assessment of criticality and informed decision-making. Regular monitoring and inspection with NDT techniques ensure ongoing evaluation of component integrity. The integration of NDT within the ECA process allows engineers to design and implement effective inspection and maintenance programs. Overall, NDT enhances the safety and reliability of cross-country pipeline installations by providing valuable information for risk mitigation and ensuring the integrity of critical components.

Coffer Dam

Cofferdams in Offshore Construction:

Ensuring Safety and Efficiency in Subsea Operations

In the realm of offshore construction work, cofferdams play a critical role in creating a temporary dry workspace in subsea environments.

These structures provide a vital means to facilitate construction, inspection, repair, or maintenance activities below the waterline. Here's a comprehensive look at the significance and utility of cofferdams:

1. Enabling Subsea Construction:

Cofferdams establish a sealed enclosure around the work area, allowing water to be pumped out, and creating a dry environment for construction activities. This enclosure prevents water ingress, creating a safe and accessible workspace for various subsea operations.

2. Supporting Diverse Applications:

Cofferdams are utilized in various offshore constructions, including the installation of underwater pipelines, foundation construction for offshore platforms, underwater welding and cutting, inspection and repair of subsea structures or equipment, and many other specialized tasks.

3. Enhancing Safety:

By isolating the work area from surrounding water, cofferdams provide a controlled environment that minimizes risks associated with tides, currents, and unpredictable subsea conditions. They ensure improved safety, allowing efficient completion of construction activities under more manageable circumstances.

4. Facilitating Efficient Construction:

Cofferdams improve work efficiency by creating a dry and stable environment. They eliminate the need for divers or costly underwater operations, allowing personnel to work comfortably and use specialized construction equipment more effectively.

5. Offering Flexibility and Versatility:

Cofferdam designs can be tailored to suit specific project requirements, considering factors such as water depth, seabed conditions, anticipated loads, and environmental conditions. They can be temporary or permanent structures, designed to withstand external pressures and with appropriate sealing mechanisms.

6. Supporting Environmentally Friendly Practices:

Cofferdam construction enables environmentally friendly practices by creating a seal that prevents the release of sediments, chemicals, or contaminants into the surrounding marine ecosystem. This promotes sustainable operations, preserving the ecosystem's integrity and complying with environmental regulations.

7. Integration with Advanced Technologies:

Cofferdams can integrate with advanced technologies such as remotely operated vehicles (ROVs), sonar systems, and remote monitoring to enhance construction operations, inspection, and surveillance activities beneath the waterline. These technologies improve precision and data collection, optimizing project outcomes.

Cofferdams provides a versatile and essential solution for offshore construction work, ensuring a controlled and safe workspace in waterlogged environments. By enabling efficient execution of subsea projects while prioritizing worker safety and environmental protection, cofferdams play a pivotal role in the success of offshore construction operations.

GOSP

GOSP stands for Gas Oil Separation Plant. It is a vital component in the oil and gas industry, specifically in upstream operations, where it plays a crucial role in separating gas, oil, and water from the produced fluid stream.

The Gas Oil Separation Plant functions as a central processing facility for the separation of the well stream into its constituent components.

Typically, the well stream consists of a mixture of crude oil, natural gas, water, and other impurities.

The GOSP employs a series of equipment and processes to carry out efficient separation. This includes various separator vessels, heaters, pumps, and other auxiliary equipment. The separator vessels, similar to those used in gas-oil separation described earlier, facilitate the separation of gas, oil, and water by utilizing differences in density.

The primary purpose of a GOSP is to separate the valuable components, namely gas, and oil, from the produced fluid. The separated gas is then processed further for purification, treatment, compression, and transportation, either to be utilized as fuel or fed into a gas processing plant.

Once the gas is separated, the remaining liquid phase, primarily crude oil, undergoes additional treatment and conditioning in the GOSP. This includes further separation of any remaining water and treatment to remove impurities, such as sediment, salts, and other contaminants.

The treated and separated oil is then directed to storage tanks or pipelines for transportation to refineries, where it will undergo further processing to convert it into various end products.

The GOSP is an essential facility in the oil and gas industry, acting as a central hub for processing, separating, and conditioning produced fluids from the wells. It enables efficient separation of gas and oil components, facilitating the extraction and utilization of valuable resources while ensuring compliance with safety, environmental, and operational requirements.

Water/Gas injection

Water injection and gas injection wells play a vital role in the oil and gas industry, particularly in the realm of enhanced oil recovery (EOR) techniques. These wells are designed and utilized to improve reservoir performance and maximize hydrocarbon production.

Water Injection Wells:

Water injection wells involve injecting water into an oil reservoir to maintain or increase pressure and displace oil towards production wells. The injected water helps maintain reservoir pressure, which gradually declines as oil is produced over time. By maintaining pressure, water injection supports reservoir drive mechanisms and aids in the recovery of additional oil that may be trapped within the reservoir.

The water injected into the reservoir can come from various sources, such as freshwater, brackish water, or even produced water that undergoes treatment. The injected water acts as a displacing fluid, pushing the oil toward production wells by creating a pressure gradient. This helps sweep and extract a higher percentage of the oil in place.

Gas Injection Wells:

Gas injection wells involve injecting gas, typically natural gas or carbon dioxide (CO_2), into the reservoir. Similar to water injection, gas injection aims to maintain reservoir pressure or increase it to improve oil recovery. The main objective of gas injection is to provide energy to the reservoir, decrease oil viscosity, and displace oil towards production wells.

Gas injection can operate through different mechanisms: immiscible displacement, miscible displacement, or a combination of both. In immiscible displacement, the injected gas, such as natural gas, does not significantly mix with the oil. It acts as a gas cap above the oil, providing pressure support and driving the oil towards the production wells. In miscible displacement, gas, usually CO_2, dissolves in the oil, reducing its viscosity and improving displacement efficiency.

Purpose of Water Injection and Gas Injection Wells:

The primary purpose of water injection and gas injection wells is to maximize oil recovery from reservoirs. These techniques enhance the displacement and sweep efficiency, helping to access and extract additional hydrocarbons that may otherwise be left behind. By maintaining or augmenting reservoir pressure, these injection methods provide the necessary energy and fluid displacement to promote efficient recovery.

Water injection and gas injection wells are vital components of EOR strategies, which aim to increase oil recovery beyond what can be achieved through primary and secondary recovery methods alone. Successful implementation of these injection techniques can significantly extend the productive life of a reservoir and increase overall production rates.

Overall, water injection and gas injection wells are instrumental in optimizing reservoir performance, sustaining production levels, and maximizing the recovery of valuable hydrocarbon resources. These techniques contribute to the efficient and sustainable exploitation of oil reservoirs while ensuring the optimal utilization of available resources.

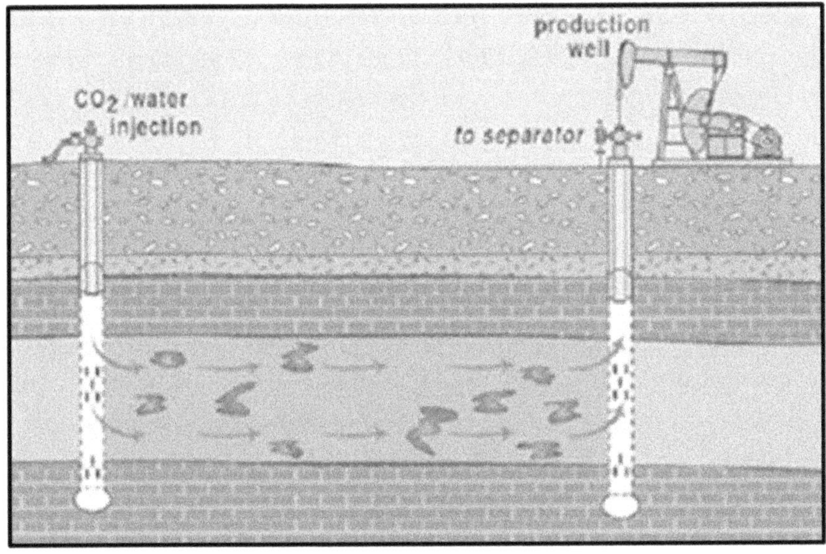

How crude oil is refined into petroleum products in a refinery

Petroleum refineries convert (refine) crude oil into petroleum products for use as fuels for transportation, heating, paving roads, and generating electricity and as feedstocks for making chemicals.

Refining breaks crude oil down into its various components, which are then selectively reconfigured into new products. Petroleum refineries are complex and expensive industrial facilities. All refineries have three basic steps:

- Separation
- Conversion
- Treatment

Separation

Modern separation involves piping crude oil through hot furnaces. The resulting liquids and vapors are discharged into distillation units. All refineries have atmospheric distillation units, but more complex refineries may have vacuum distillation units.

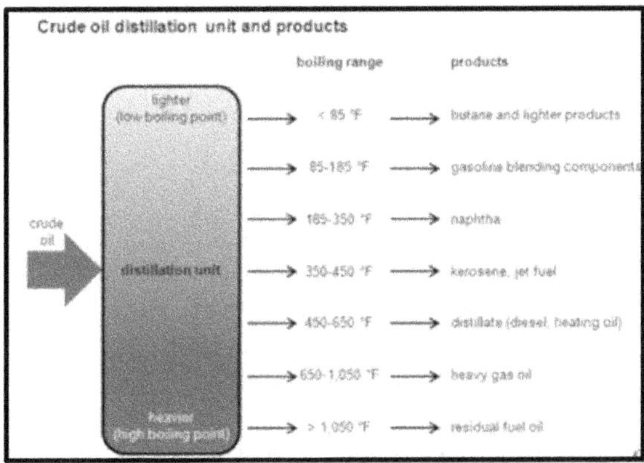

Inside the distillation units, the liquids and vapors separate into petroleum components, called *fractions*, according to their boiling points. Heavy fractions are on the bottom and light fractions are on the top.

The lightest fractions, including gasoline and liquefied refinery gases, vaporize and rise to the top of the distillation tower, where they condense back to liquids.

Medium-weight liquids, including kerosene and distillates, stay in the middle of the distillation tower.

Heavier liquids, called gas oils, separate lower down in the distillation tower, and the heaviest fractions with the highest boiling points settle at the bottom of the tower.

Conversion

After distillation, heavy, lower-value distillation fractions can be processed further into lighter, higher-value products such as gasoline. At this point in the process, fractions from the distillation units are transformed into *streams* (intermediate components) that eventually become finished products.

The most widely used conversion method is called *cracking* because it uses heat, pressure, catalysts, and sometimes hydrogen to crack heavy hydrocarbon molecules into lighter ones. A cracking unit consists of one or more tall, thick-walled, rocket-shaped reactors and a network of furnaces, heat exchangers, and other vessels. Complex refineries may have one or more types of crackers, including fluid catalytic cracking units and hydrocracking/hydrocracker units.

Cracking is not the only form of crude oil conversion. Other refinery processes rearrange molecules rather than splitting molecules to add value.

Alkylation, for example, makes gasoline components by combining some of the gaseous byproducts of cracking. The process, which essentially is cracking in reverse, takes place in a series of large, horizontal vessels and tall, skinny towers.

Reforming uses heat, moderate pressure, and catalysts to turn naphtha, a light, relatively low-value fraction, into high-octane gasoline components.

Treatment

The finishing touches occur during the final treatment. To make gasoline, refinery technicians carefully combine a variety of streams from the processing units. Octane level, vapor pressure ratings, and other special considerations determine the gasoline blend.

Storage

Both incoming crude oil and the outgoing final products are stored temporarily in large tanks on a tank farm near the refinery. Pipelines, trains, and trucks carry the final products from the storage tanks to locations across the country.

Refined Crude Oil Product	Share of Crude Oil Refined
Gasoline	42.7%
Diesel	27.4%
Jet fuel	5.8%
Heavy fuel	5.0%
Asphalt	4.0%
Light fuel	3.0%
Hydrocarbon gas liquids	2.0%
Other	10.1%

Crude oil not only powers our vehicles but also helps pave the roads we drive on. About 4% of refined crude oil becomes asphalt, which is used to make concrete and different kinds of sealing and insulation products.

Although transportation and utility fuels dominate a large proportion of refined products, essential everyday materials like wax and plastic are also dependent on crude oil. With about 10% of refined products used to make plastics, cosmetics, and textiles, a barrel of crude oil can produce a variety of unexpected everyday products.

Personal care products like cosmetics and shampoo are made using petroleum products, as are medical supplies like IV bags and pharmaceuticals. Modern life would look very different without crude oil.

This means the feedstock for the petrochemical industry is also Crude oil.

You can now imagine, the importance of crude oil in our life. That is the exact reason why it is called liquid gold.

Now let us delve into the petrochemical industry.......

Petrochemical industry

The petrochemicals industry has evolved out of oil and gas processing by adding value to low-value by-products, which have limited use in the fuels industry. The industry now produces a remarkable range of useful products, including plastics, synthetic rubber, solvents, fertilizers, pharmaceuticals, additives, explosives, and adhesives. These materials have important applications in almost all areas of modern society. Petrochemical products are used in cars, packaging, household goods, medical equipment, paints, clothing, and building materials to name just a few of the common applications. Furthermore, the industry continues to innovate through new technology and the ability to process different types of raw materials.

The petrochemicals industry sources raw materials from refining and gas processing and converts these raw materials into valuable products using a variety of chemical process technologies. A variety of feedstocks are used as raw materials, the industry driver being economic. If cheap feedstock is available, then there will always be somebody prepared to try to make a profit by making something valuable out of it. These feedstocks are subject to a variety of processes to produce a handful of chemical building blocks. These building blocks are further processed through a variety of reactions to form the final petrochemical products.

The target markets for petrochemical products are smaller and more specialized in comparison to refined products and natural gas. Although petrochemical products usually earn premium prices compared to refined products and natural gas products, marketing is more demanding. Market risks and competitive analysis therefore play an important role in lender assessment of petrochemical project finance transactions.

Liquified Natural Gas (LNG)

Natural gas is extracted from subsurface rock formations via drilling. Advances in hydraulic fracturing technologies have enabled access to large volumes of natural gas from shale.

To use the LNG, it must be reheated to turn it back into its gaseous form. This process is called re-gasification. In 95% of locations, seawater vaporizers are used to raise the temperature of the liquid to turn it back into a gas

LNG is natural gas that has been cooled to −260° F (−162° C), changing it from a gas into a liquid that is 1/600th of its original volume. This dramatic reduction allows it to be shipped safely and efficiently aboard specially designed LNG vessels.

COMPRESSED NATURAL GAS VS. LIQUEFIED NATURAL GAS

Compressed Natural Gas, or CNG, and Liquefied Natural Gas, or LNG, are the same substance. CNG is received and stored in a vehicle's tank in gaseous form. To obtain LNG, natural gas is compressed and cooled to extremely low temperatures, at which point it turns to liquid. LNG can then be shipped, stored, and used to fill the tanks of LNG vehicles. Much of the global natural gas trade occurs in the form of LNG. Some countries, such as South Korea and Japan, receive almost all of the natural gas they use in LNG form.

In vehicle applications, the main advantage that LNG has over CNG is that it is denser. For two tanks of the same size, the LNG tank will allow a vehicle to drive further than the CNG tank. This makes LNG an interesting option for heavy trucks traveling long distances.

Renewable energy

Renewable energy is energy derived from natural sources that are replenished at a higher rate than they are consumed. Sunlight and wind, for example, are such sources that are constantly being replenished. Renewable energy sources are plentiful and all around us.

Fossil fuels - coal, oil, and gas - on the other hand, are non-renewable resources that take hundreds of millions of years to form. Fossil fuels, when burned to produce energy, cause harmful greenhouse gas emissions, such as carbon dioxide.

Generating renewable energy creates far lower emissions than burning fossil fuels. Transitioning from fossil fuels, which currently account for the lion's share of emissions, to renewable energy is key to addressing the climate crisis.

Renewables are now cheaper in most countries and generate three times more jobs than fossil fuels

Types of renewable energy

Solar energy

Solar energy is the most abundant of all energy resources and can even be harnessed in cloudy weather. The rate at which solar energy is intercepted by the Earth is about 10,000 times greater than the rate at which humankind consumes energy.

Solar technologies can deliver heat, cooling, natural lighting, electricity, and fuels for a host of applications. Solar technologies convert sunlight into electrical energy either through photovoltaic panels or through mirrors that concentrate solar radiation.

Although not all countries are equally endowed with solar energy, a significant contribution to the energy mix from direct solar energy is possible for every country.

The cost of manufacturing solar panels has plummeted dramatically in the last decade, making them not only affordable but often the cheapest form of electricity. Solar panels have a lifespan of roughly 30 years and come in a variety of shades depending on the type of material used in manufacturing.

Wind energy

Wind energy harnesses the kinetic energy of moving air by using large wind turbines located on land (onshore) or in the sea- or freshwater (offshore). Wind energy has been used for millennia, but onshore and offshore wind energy technologies have evolved over the last few years to maximize the electricity produced - with taller turbines and larger rotor diameters.

Though average wind speeds vary considerably by location, the world's technical potential for wind energy exceeds global electricity production, and ample potential exists in most regions of the world to enable significant wind energy deployment.

Many parts of the world have strong wind speeds, but the best locations for generating wind power are sometimes remote ones. Offshore wind power offers tremendous potential.

Geothermal energy

Geothermal energy utilizes the accessible thermal energy from the Earth's interior. Heat is extracted from geothermal reservoirs using wells or other means.

Reservoirs that are naturally sufficiently hot and permeable are called hydrothermal reservoirs, whereas reservoirs that are sufficiently hot but that are improved with hydraulic stimulation are called enhanced geothermal systems.

Once at the surface, fluids of various temperatures can be used to generate electricity. The technology for electricity generation from hydrothermal reservoirs is mature and reliable and has been operating for more than 100 years.

Hydropower

Hydropower harnesses the energy of water moving from higher to lower elevations. It can be generated from reservoirs and rivers. Reservoir hydropower plants rely on stored water in a reservoir, while run-of-river hydropower plants harness energy from the available flow of the river.

Hydropower reservoirs often have multiple uses - providing drinking water, water for irrigation, flood and drought control, navigation services, as well as energy supply.

Hydropower currently is the largest source of renewable energy in the electricity sector. It relies on generally stable rainfall patterns and can be negatively impacted by climate-induced droughts or changes to ecosystems that impact rainfall patterns.

The infrastructure needed to create hydropower can also impact ecosystems in adverse ways. For this reason, many consider small-scale hydro a more environmentally-friendly option, and especially suitable for communities in remote locations.

Ocean energy

Ocean energy derives from technologies that use the kinetic and thermal energy of seawater - waves or currents for instance - to produce electricity or heat.

Ocean energy systems are still at an early stage of development, with several prototype wave and tidal current devices being explored. The theoretical potential for ocean energy easily exceeds present human energy requirements.

Bioenergy

Bioenergy is produced from a variety of organic materials, called biomass, such as wood, charcoal, dung, and other manures for heat and power production, and crops for liquid biofuels. Most biomass is used in rural areas for cooking, lighting, and space heating, generally by poorer populations in developing countries.

Modern biomass systems include dedicated crops or trees, residues from agriculture and forestry, and various organic waste streams.

The energy created by burning biomass creates greenhouse gas emissions but at lower levels than burning fossil fuels like coal, oil, or gas. However, bioenergy should only be used in limited applications, given the potential negative environmental impacts related to large-scale increases in forest and bioenergy plantations, and resulting deforestation and land-use change.

Sustainable energy

Each year the stakes grow higher in the fight to save the environment and combat global warming. Now more than ever, we're aware of the damaging effects that our current dependence on fossil fuels holds for our collective futures. One of the key solutions? Sustainable energy.

Fossil fuels (e.g., coal, natural gas, and oil) are not only harmful to the planet when burned daily for energy, but they're also unsustainable as finite resources. Sustainability refers to the concept that all people can meet their basic needs infinitely, without compromising future generations. Sustainability in terms of energy embraces the same principles.

One day the world may run out of fossil fuels.

Difference between sustainable and renewable energy?

People often use the terms "sustainable" and "renewable" interchangeably. However, there is a difference between the two: the possibility of replenishment.

Sustainable energy, as highlighted above, is theoretically inexhaustible. It cannot be depleted because sustainable energy sources don't need to be replenished. For example, think of the sun or wind. Neither resource needs to be created or replaced.

On the other hand, renewable energy is theoretically exhaustible — it uses resources from the earth that can naturally be replenished, such as crops and biomatter. A renewable energy source like bioenergy uses biological masses (e.g., agricultural byproducts like straw and manure) to create energy. Another example of bioenergy is ethanol, which is made from sugarcane and corn. Since these crops can be planted and farmed to generate more energy, it's a type of renewable energy.

Clean energy

What is the role of clean energy transitions?

Clean hydrogen produced with renewable or nuclear energy, or fossil fuels using carbon capture, can help to decarbonize a range of sectors, including long-haul transport, chemicals, and iron and steel, where it has proven difficult to reduce emissions. Hydrogen-powered vehicles would improve air quality and promote energy security. Hydrogen can also support the integration of variable renewables in the electricity system, being one of the few options for storing energy over days, weeks, or months.

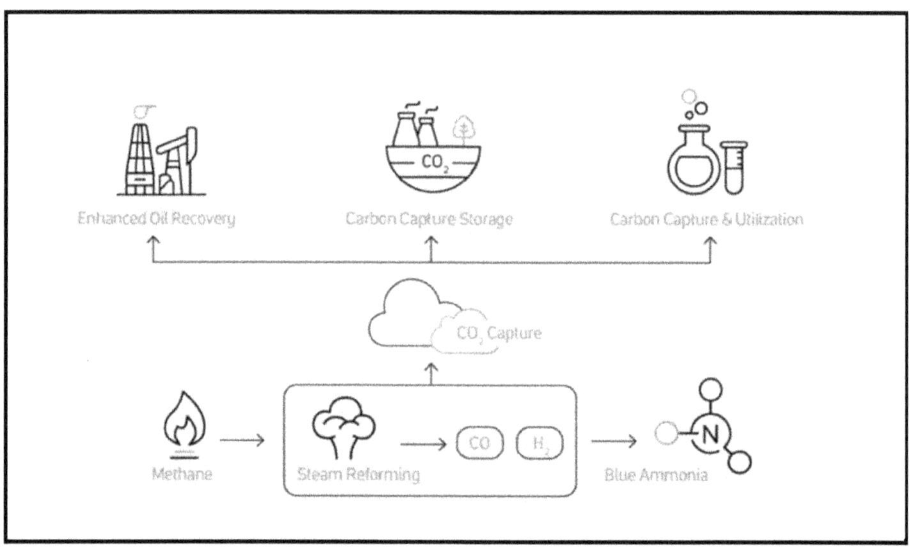

TITBITS

Difference between Inquiry and enquiry

The difference between inquiry and enquiry is minor and deals with nuance in meaning: inquiry is preferred for formal requests and official investigations. enquiry is much broader, referring to any requests, formal or informal.

Difference between Advice and Advise

Advice is a noun that refers to an opinion or suggestion that is given. It's pronounced with an "s" sound at the end.

Advise is a verb that refers to the act of giving an opinion or suggestion. It's pronounced with a "z" sound at the end (though it is never spelled "advize")

TITBITS

Communication skills

1. **Active Listening**: Effective communication starts with listening. Pay attention to the speaker, show empathy, and avoid interrupting. (Keep the mobile phone away)
2. **Clear and Concise**: Keep your messages clear and concise to avoid confusion. Avoid unnecessary jargon and be direct in your communication.
3. **Non-Verbal Communication**: Body language, facial expressions, and eye contact can convey a lot of information. Be aware of your non-verbal cues and ensure they align with your words.

4. **Ask Questions**: Asking questions not only shows that you are engaged, but it also ensures that you understood the message correctly.

5. **Be Open and Respectful**: Respect others' opinions and be open to feedback. This can help build trust and foster better communication.

6. **Use Appropriate Tone and Language**: The tone and language should be appropriate for the situation and the audience. What works in a casual conversation with friends might not work in a professional setting.

7. **Practice Empathy**: Try to understand the other person's perspective. This can help you communicate more effectively and build stronger relationships.

8. **Written Communication**: For written communication, ensure your messages are well-structured and free of grammatical errors. Use tools to check spelling and grammar if necessary.

Communication is a two-way process and involves both speaking and listening. Practice these skills regularly to become a better communicator.

I have written one section, on *technical writing, (page 307)* which will be of help for engineers while preparing the procedures, technical queries, and other documents.

CHAPTER 5

Engineering

Before getting into the details of the Detailed Engineering, let us understand the platform.

3D modeling

The 3D model is the heart of detailed engineering around which all activities take place.

An engineer needs to get familiarized with the 3D modeling concept, irrespective of the modeling software, which is very essential.

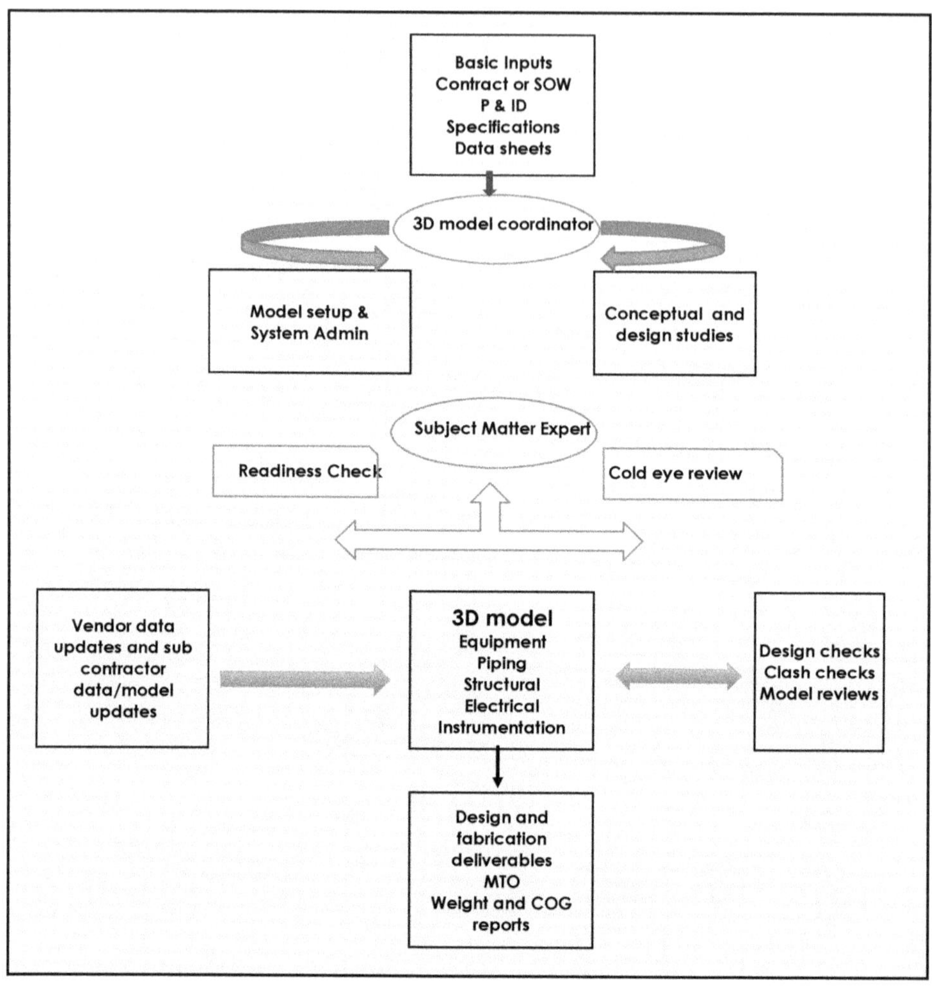
Basic Inputs
Contract or SOW
P & ID
Specifications
Data sheets
3D model coordinator
Model setup &
System Admin
Conceptual and
design studies
Subject Matter Expert
Readiness Check
Cold eye review
Vendor data
updates and sub
contractor
data/model
updates
3D model
Equipment
Piping
Structural
Electrical
Instrumentation
Design checks
Clash checks
Model reviews
Design and
fabrication
deliverables
MTO
Weight and COG
reports

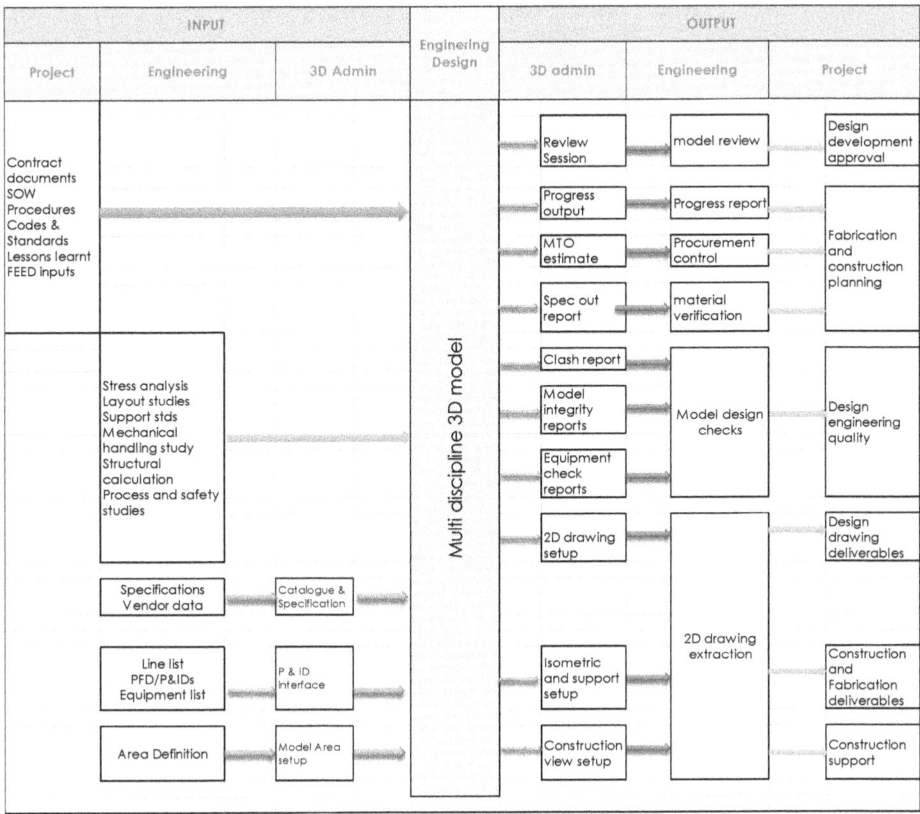

In the oil and gas industry, the use of 3D modeling in engineering design has become increasingly prevalent, revolutionizing the way plants are designed and constructed. A 3D model, also known as a digital twin or virtual representation, refers to a comprehensive digital representation of a facility or plant.

1. Enhanced Visualization and Collaboration: 3D modeling allows design teams to create a visual representation of the oil and gas plant, providing a realistic depiction of how the various components and systems will interact. This enables stakeholders to visualize the design more accurately and identify potential clashes, interferences, or design issues early in the project lifecycle. Through enhanced visualization, multidisciplinary teams

can collaborate effectively, leading to better decision-making, coordination, and improved project outcomes.

2. Design Optimization and Efficiency: 3D modeling facilitates design optimization and efficiency by simulating system performance and evaluating different design alternatives. It allows engineers to model and analyze complex processes, identify potential bottlenecks, optimize layout and spacing, and assess overall plant performance. With the ability to perform virtual tests and simulations, engineers can refine designs, minimize risks, and enhance the efficiency of oil and gas plant operations.

3. Clash Detection and Conflict Resolution: A major advantage of 3D modeling is the ability to detect clashes or conflicts between different components or systems before construction. Clash detection tools can identify clashes between structural elements, equipment, pipelines, or cables, providing an opportunity to resolve clashes and conflicts during the design phase. Early clash detection prevents costly rework and delays during construction, streamlining the project timeline and reducing overall costs.

4. Construction Planning and Sequencing: 3D models assist in construction planning and sequencing by providing a detailed representation of the oil and gas plant. This allows project teams to visualize and streamline construction sequences, optimize workflows, and plan for efficient resource allocation. 3D models serve as valuable references for contractors and construction teams, enhancing communication and coordination on the construction site.

5. Operations and Maintenance: Beyond the design and construction phases, 3D models in engineering design have long-term benefits for operations and maintenance. The accurate and detailed 3D model serves as a valuable reference for facility operators and maintenance teams, aiding in troubleshooting, maintenance planning, and future modifications and enhancements.

The use of 3D modeling in engineering design for oil and gas plants enhances collaboration, improves design optimization, minimizes clashes, and enhances construction planning and operations. With its ability to provide a comprehensive digital representation of the facility, 3D modeling is a powerful tool that drives efficiency, cost savings, and effective decision-making throughout the project lifecycle.

TITBITS

Ikigai:

Ikigai is often described as the intersection of four fundamental elements: what you love, what you are good at, what the world needs, and what you can be paid for. It is believed that finding balance and harmony among these elements leads to a sense of purpose, fulfillment, and deep life satisfaction.

To discover your ikigai, one must embark on a journey of self-discovery and introspection. This involves reflecting on personal passions, talents, and values, as well as considering how one's skills and passions can be used to contribute positively to society. By aligning these aspects, individuals can find clarity and a sense of direction, enabling them to live a meaningful life filled with joy and fulfillment.

- Identify your passions: Reflect on activities, hobbies, and interests that bring you joy and fulfillment.
- Discover your strengths: Recognize your unique talents, skills, and abilities.
- Find meaning and purpose: Explore how your passions and strengths can be aligned with the needs of the world around you.
- Seek growth opportunities: Continuously challenge yourself and engage in activities that allow you to develop and grow.

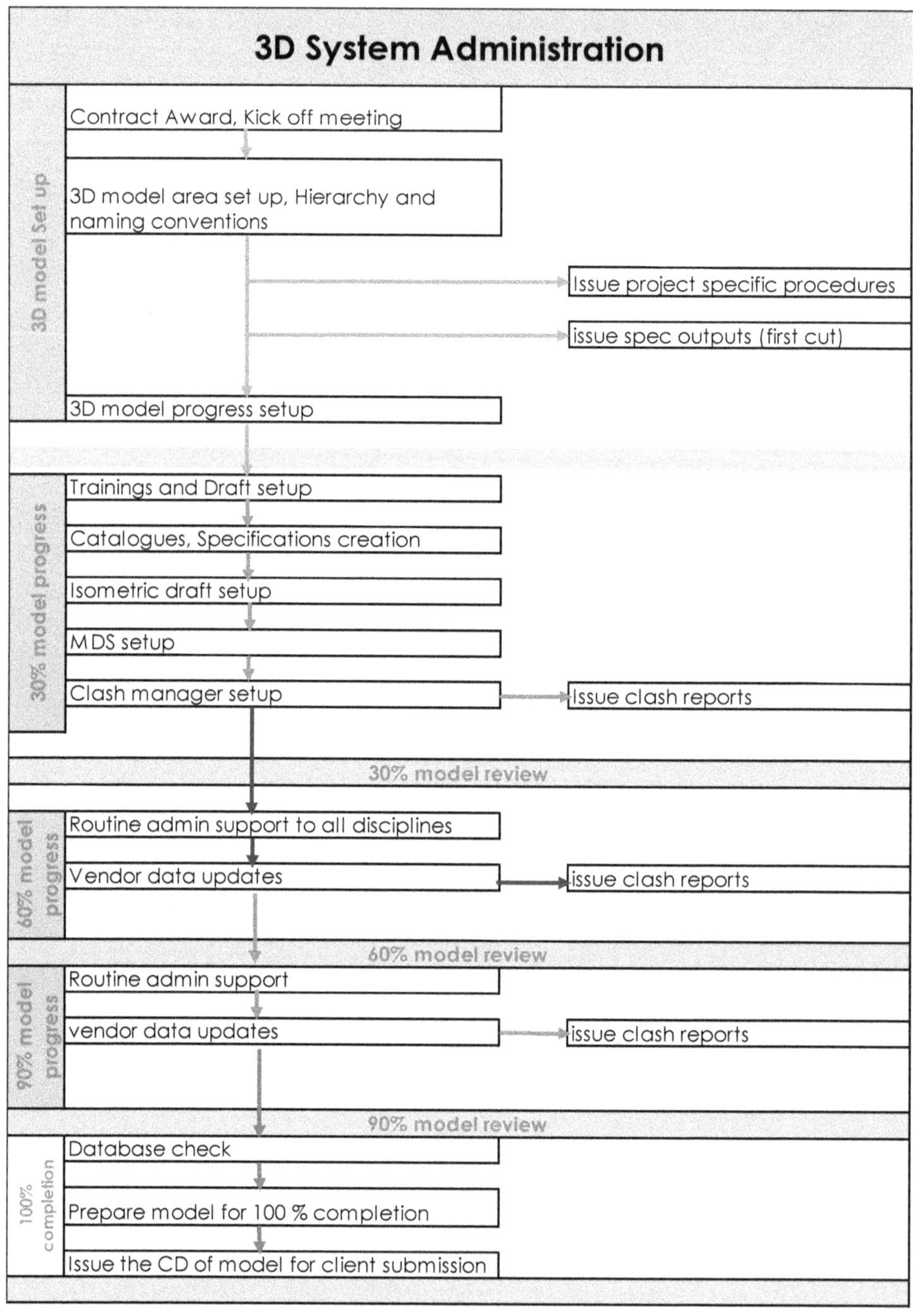

3D System Administration
Contract Award, Kick off meeting
3D model area set up, Hierarchy and naming conventions
Issue project specific procedures
issue spec outputs (first cut)
3D model progress setup
Trainings and Draft setup
Catalogues, Specifications creation
Isometric draft setup
MDS setup
Clash manager setup
Issue clash reports
30% model review
Routine admin support to all disciplines
Vendor data updates
issue clash reports
60% model review
Routine admin support
vendor data updates
issue clash reports
90% model review
Database check
Prepare model for 100 % completion
Issue the CD of model for client submission
3D model Set up
30% model progress
60% model progress
90% model progress
100% completion

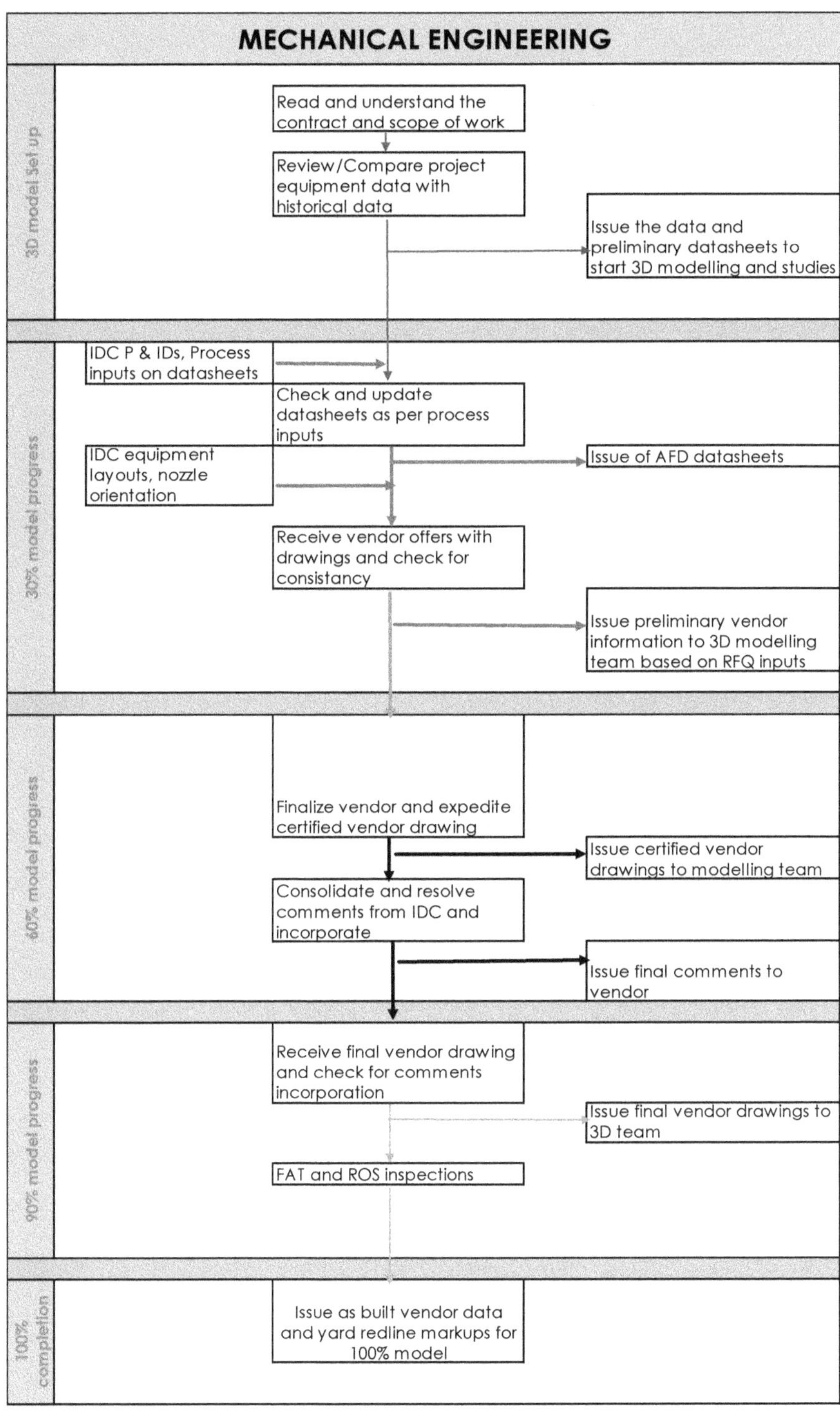

MECHANICAL ENGINEERING
3D model set up
Read and understand the contract and scope of work
Review/Compare project equipment data with historical data
Issue the data and preliminary datasheets to start 3D modelling and studies
30% model progress
IDC P & IDs, Process inputs on datasheets
Check and update datasheets as per process inputs
IDC equipment layouts, nozzle orientation
Issue of AFD datasheets
Receive vendor offers with drawings and check for consistancy
Issue preliminary vendor information to 3D modelling team based on RFQ inputs
60% model progress
Finalize vendor and expedite certified vendor drawing
Issue certified vendor drawings to modelling team
Consolidate and resolve comments from IDC and incorporate
Issue final comments to vendor
90% model progress
Receive final vendor drawing and check for comments incorporation
Issue final vendor drawings to 3D team
FAT and ROS inspections
100% completion
Issue as built vendor data and yard redline markups for 100% model

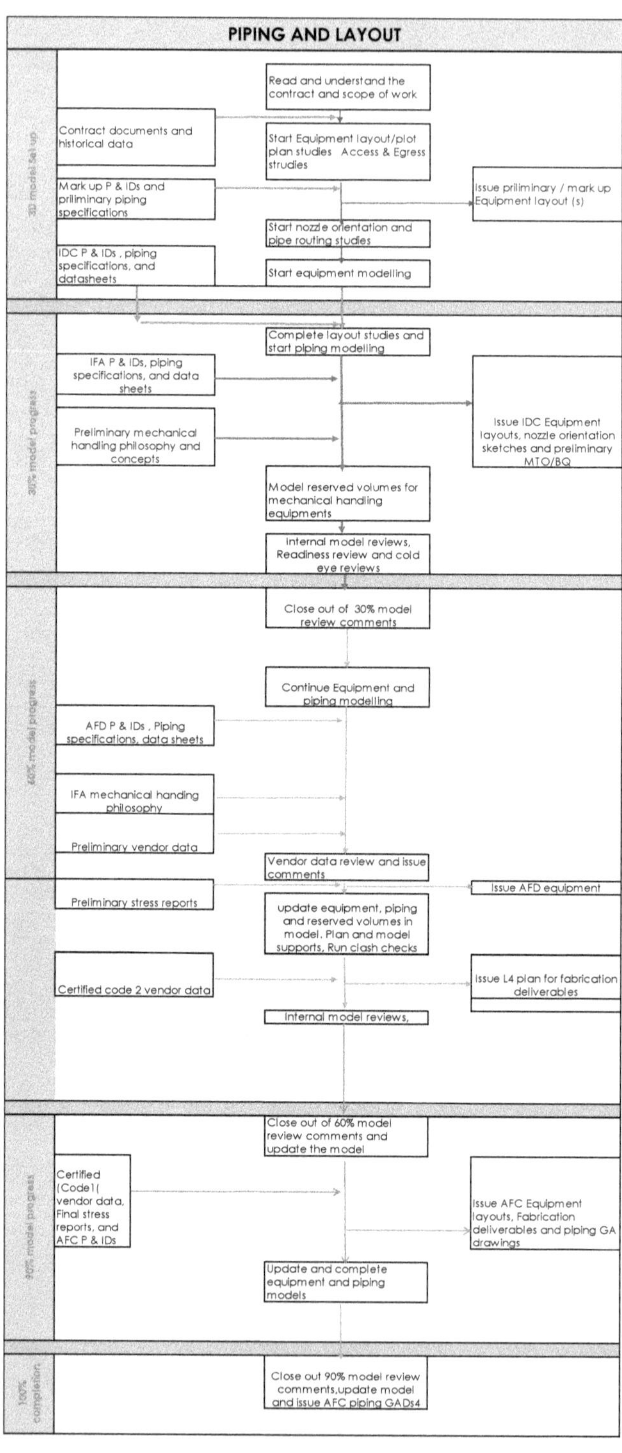

PIPING AND LAYOUT

3D model set up

Read and understand the contract and scope of work

Contract documents and historical data

Start Equipment layout/plot plan studies   Access & Egress strudies

Mark up P & IDs and priliminary piping specifications

Issue priliminary / mark up Equipment layout (s)

Start nozzle orientation and pipe routing studies

IDC P & IDs , piping specifications, and datasheets

Start equipment modelling

30% model progress

Complete layout studies and start piping modelling

IFA P & IDs, piping specifications, and data sheets

Preliminary mechanical handling philosophy and concepts

Issue IDC Equipment layouts, nozzle orientation sketches and preliminary MTO/BQ

Model reserved volumes for mechanical handling equipments

Internal model reviews, Readiness review and cold eye reviews

60% model progress

Close out of 30% model review comments

Continue Equipment and piping modelling

AFD P & IDs , Piping specifications, data sheets

IFA mechanical handing philosophy

Preliminary vendor data

Vendor data review and issue comments

Preliminary stress reports

Issue AFD equipment

update equipment, piping and reserved volumes in model. Plan and model supports, Run clash checks

Certified code 2 vendor data

Issue L4 plan for fabrication deliverables

Internal model reviews,

90% model progress

Close out of 60% model review comments and update the model

Certified (Code1(vendor data, Final stress reports, and AFC P & IDs

Issue AFC Equipment layouts, Fabrication deliverables and piping GA drawings

Update and complete equipment and piping models

100% completion

Close out 90% model review comments,update model and issue AFC piping GADs4

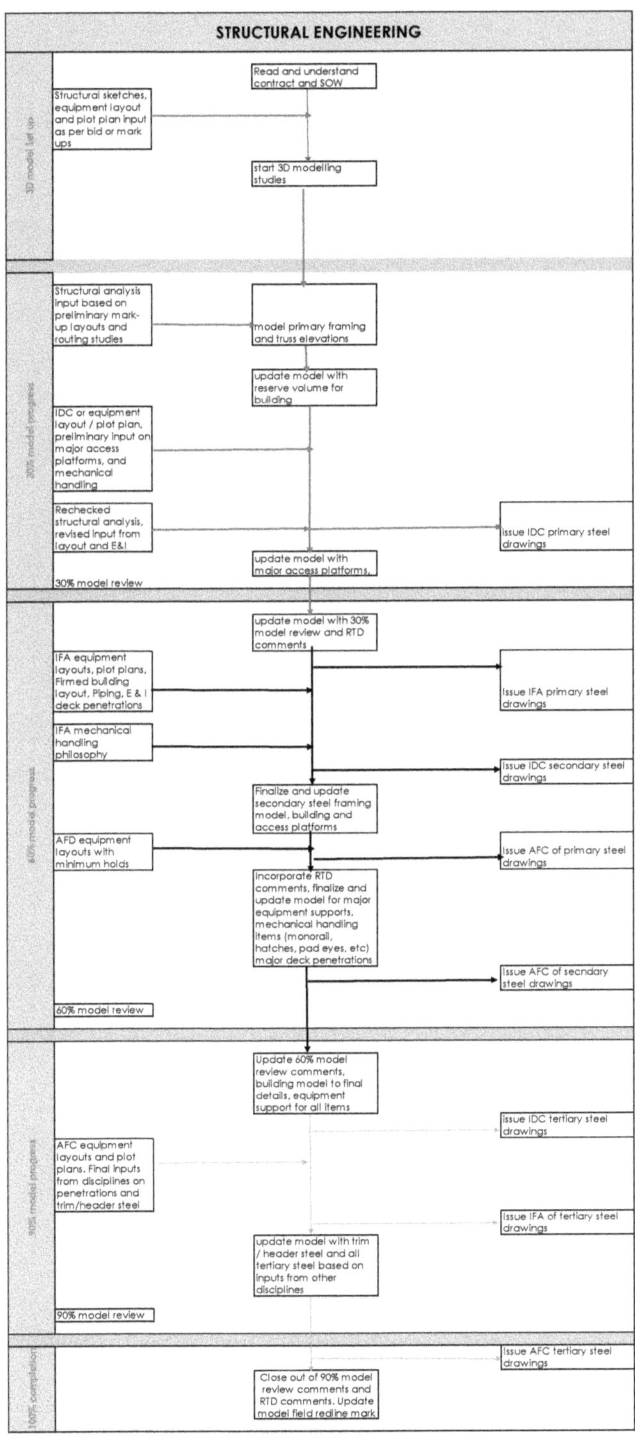
STRUCTURAL ENGINEERING
3D model set up
Structural sketches, equipment layout and plot plan input as per bid or mark ups
Read and understand contract and SOW
start 3D modelling studies
30% model progress
Structural analysis input based on preliminary mark-up layouts and routing studies
model primary framing and truss elevations
update model with reserve volume for building
IDC or equipment layout / plot plan, preliminary input on major access platforms, and mechanical handling
Rechecked structural analysis, revised input from layout and E&I
issue IDC primary steel drawings
update model with major access platforms,
30% model review
update model with 30% model review and RTD comments
IFA equipment layouts, plot plans, Firmed building layout, Piping, E & I deck penetrations
Issue IFA primary steel drawings
IFA mechanical handling philosophy
Issue IDC secondary steel drawings
60% model progress
Finalize and update secondary steel framing model, building and access platforms
AFD equipment layouts with minimum holds
Issue AFC of primary steel drawings
Incorporate RTD comments, finalize and update model for major equipment supports, mechanical handling items (monorail, hatches, pad eyes, etc) major deck penetrations
Issue AFC of secndary steel drawings
60% model review
Update 60% model review comments, building model to final details, equipment support for all items
issue IDC tertiary steel drawings
90% model progress
AFC equipment layouts and plot plans. Final inputs from disciplines on penetrations and trim/header steel
Issue IFA of tertiary steel drawings
update model with trim / header steel and all tertiary steel based on inputs from other disciplines
90% model review
100% completion
Issue AFC tertiary steel drawings
Close out of 90% model review comments and RTD comments. Update model field redline mark

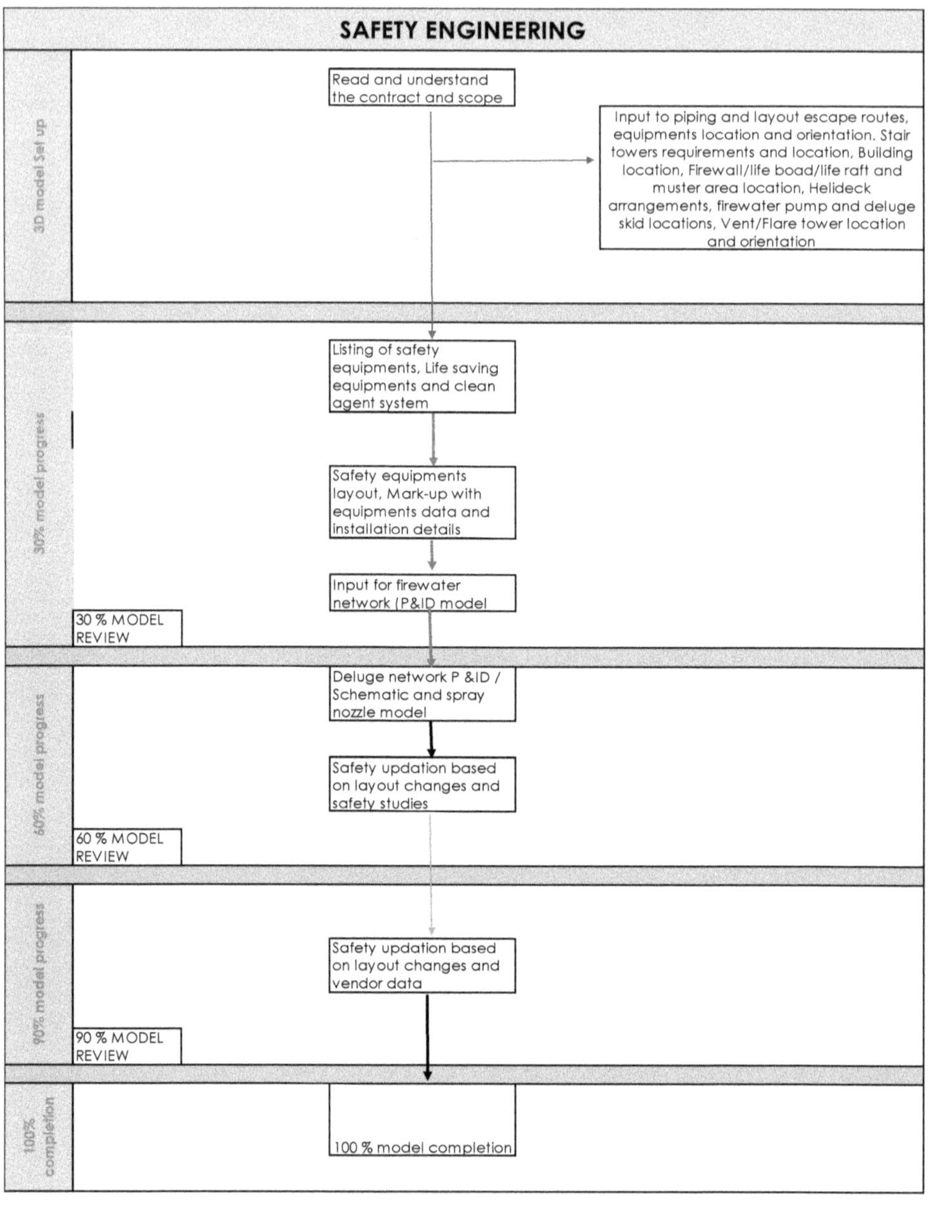
SAFETY ENGINEERING

3D model Set up

Read and understand the contract and scope

Input to piping and layout escape routes, equipments location and orientation. Stair towers requirements and location, Building location, Firewall/life boad/life raft and muster area location, Helideck arrangements, firewater pump and deluge skid locations, Vent/Flare tower location and orientation

30% model progress

Listing of safety equipments, Life saving equipments and clean agent system

Safety equipments layout, Mark-up with equipments data and installation details

Input for firewater network (P&ID model

30 % MODEL REVIEW

60% model progress

Deluge network P &ID / Schematic and spray nozzle model

Safety updation based on layout changes and safety studies

60 % MODEL REVIEW

90% model progress

Safety updation based on layout changes and vendor data

90 % MODEL REVIEW

100% completion

100 % model completion

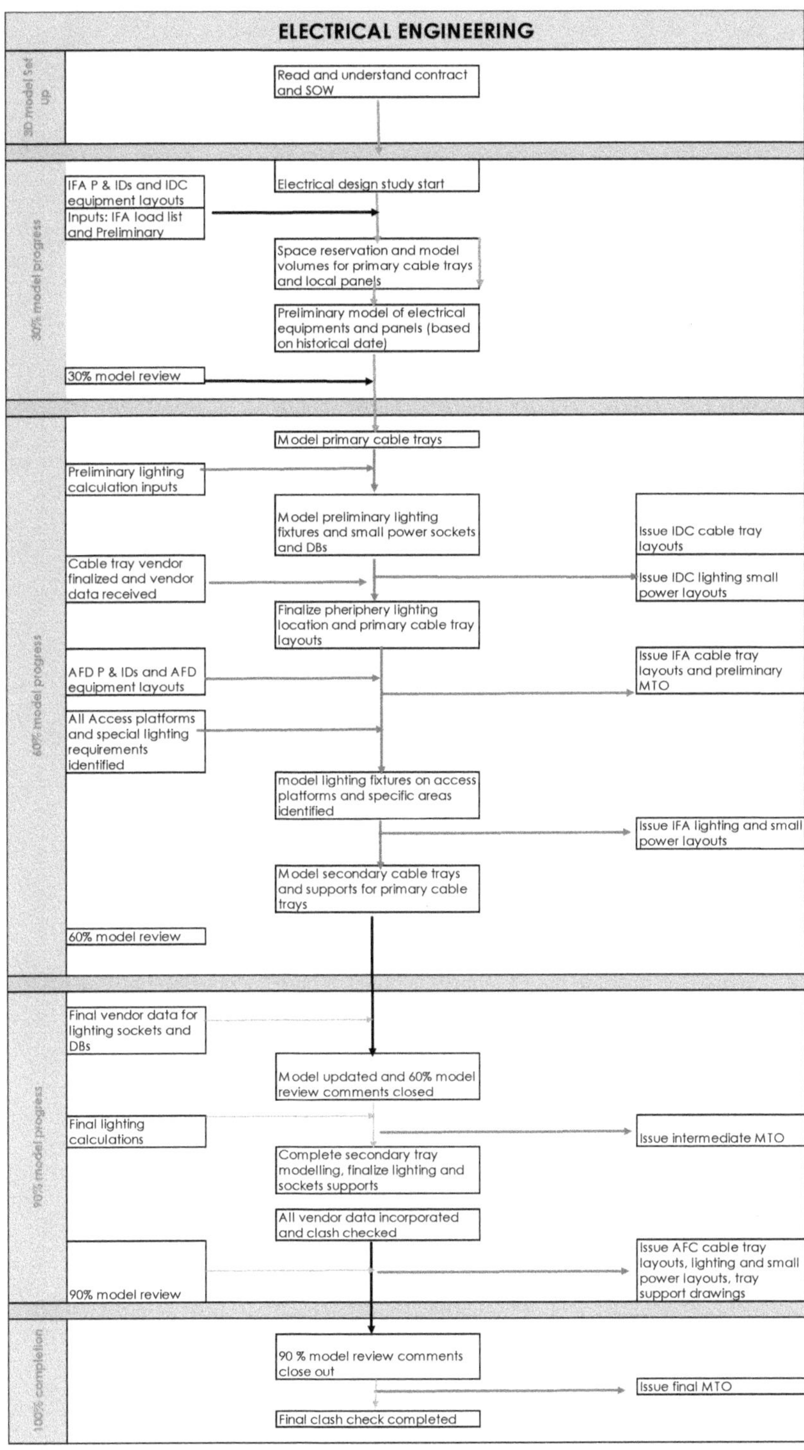
ELECTRICAL ENGINEERING
3D model Set up
30% model progress
60% model progress
90% model progress
100% completion
Read and understand contract and SOW
IFA P & IDs and IDC equipment layouts
Inputs: IFA load list and Preliminary
Electrical design study start
Space reservation and model volumes for primary cable trays and local panels
Preliminary model of electrical equipments and panels (based on historical date)
30% model review
Model primary cable trays
Preliminary lighting calculation inputs
Model preliminary lighting fixtures and small power sockets and DBs
Issue IDC cable tray layouts
Cable tray vendor finalized and vendor data received
Issue IDC lighting small power layouts
Finalize pheriphery lighting location and primary cable tray layouts
AFD P & IDs and AFD equipment layouts
Issue IFA cable tray layouts and preliminary MTO
All Access platforms and special lighting requirements identified
model lighting fixtures on access platforms and specific areas identified
Issue IFA lighting and small power layouts
Model secondary cable trays and supports for primary cable trays
60% model review
Final vendor data for lighting sockets and DBs
Model updated and 60% model review comments closed
Final lighting calculations
Issue intermediate MTO
Complete secondary tray modelling, finalize lighting and sockets supports
All vendor data incorporated and clash checked
Issue AFC cable tray layouts, lighting and small power layouts, tray support drawings
90% model review
90 % model review comments close out
Issue final MTO
Final clash check completed

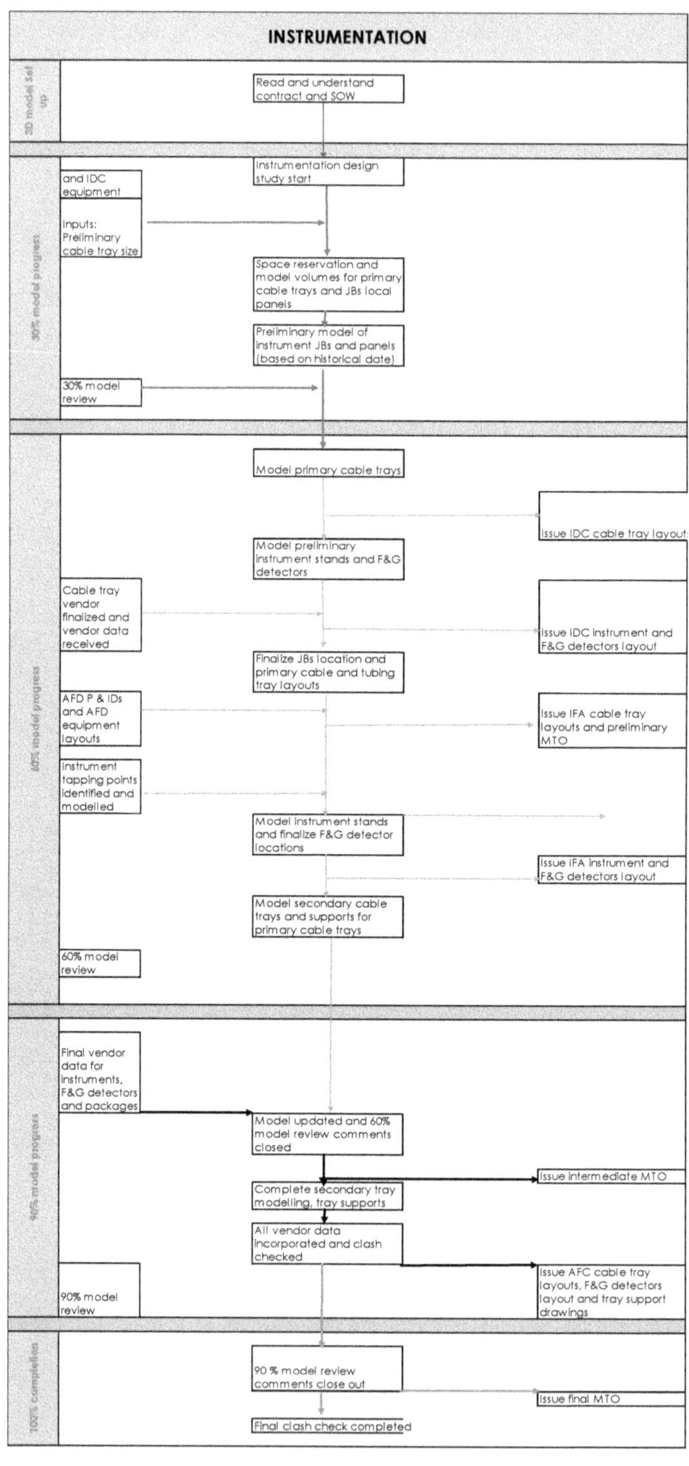
INSTRUMENTATION
3D model Set up
Read and understand contract and SOW
30% model progress
and IDC equipment
Inputs: Preliminary cable tray size
Instrumentation design study start
Space reservation and model volumes for primary cable trays and JBs local panels
Preliminary model of instrument JBs and panels (based on historical date)
30% model review
60% model progress
Model primary cable trays
Issue IDC cable tray layout
Model preliminary instrument stands and F&G detectors
Cable tray vendor finalized and vendor data received
Issue IDC instrument and F&G detectors layout
Finalize JBs location and primary cable and tubing tray layouts
AFD P & IDs and AFD equipment layouts
Issue IFA cable tray layouts and preliminary MTO
Instrument tapping points identified and modelled
Model instrument stands and finalize F&G detector locations
Issue IFA instrument and F&G detectors layout
Model secondary cable trays and supports for primary cable trays
60% model review
90% model progress
Final vendor data for instruments, F&G detectors and packages
Model updated and 60% model review comments closed
Issue intermediate MTO
Complete secondary tray modelling, tray supports
All vendor data incorporated and clash checked
Issue AFC cable tray layouts, F&G detectors layout and tray support drawings
90% model review
100% completion
90 % model review comments close out
Issue final MTO
Final clash check completed

Engineering disciplines

In detailed design engineering for the oil and gas sector, various engineering disciplines collaborate and interface to ensure a comprehensive and efficient design. These disciplines include:

1. Process Engineering:

Process engineering focuses on designing the optimal refining or production processes for oil and gas operations. It involves determining the required equipment, specifications, and operating conditions for each unit within the facility. Process engineers work closely with other disciplines to ensure the compatibility of equipment and the efficient utilization of resources.

2. Mechanical Engineering:

Mechanical engineering encompasses the design and selection of equipment, machinery, and piping systems. Mechanical engineers collaborate with process engineers to ensure that the equipment meets the process requirements and operates reliably and safely. They also consider factors such as material selection, heat transfer, pressure drop, and mechanical stress analysis in the design process. Mechanical engineering is divided into three categories, viz. Static, Rotating, and Packaged equipments.

3. Civil and Structural Engineering:

Civil and structural engineers design the foundations, structures, and support systems for various components within the oil and gas facilities. Their work includes determining the load-bearing capacities, seismic considerations, and environmental factors to ensure the structural integrity of the facility. They also collaborate with other disciplines to ensure the safe installation and integration of equipment.

4. Electrical Engineering:

Electrical engineering involves the design and integration of electrical systems, including power distribution, lighting, control systems, and instrumentation. Electrical engineers work closely with mechanical and process engineers to ensure the compatibility and safety of the electrical systems (Hazardous area classifications) within the facility. They also consider factors such as load calculations, electrical codes, and standards in their designs.

5. Instrumentation and Control Engineering:

Instrumentation and control engineering focuses on the design and implementation of control systems, instrumentation, and automation technologies. Instrumentation engineers collaborate with process and electrical engineers to ensure the accurate measurement, control, and monitoring of various process parameters within the facility. They also interface with software engineers to develop efficient and reliable control algorithms.

6. Piping and Layout Engineering

Piping and Layout engineering form the center of 3D modeling in a typical project.

This engineering discipline consists of:

1. Piping engineering
2. Stress analysis
3. Materials
4. Layout design
5. 3D modeling

Piping design encompasses the selection of appropriate pipe materials, sizing, and routing of pipes based on the fluid properties, pressure, temperature, and flow requirements. Design considerations also include the prevention of leaks, and corrosion, and the accommodation of thermal expansion and contraction.

Equipment Layout: Piping layout engineers work closely with other disciplines, like process, mechanical, and structural engineering, to determine the optimal layout for components, including vessels, pumps, heat exchangers, compressors, and control valves. The equipment layout must consider accessibility for operation, maintenance, and safety requirements.

Plot Plan Development: Plot plan development is a crucial component of piping and layout engineering. It involves establishing the optimal location and orientation of equipment, structures, and piping networks within the facility. Considerations for plot plan development include space requirements, process flow optimization, operational efficiency, and adherence to safety regulations.

Pipe Routing and Support Design: Pipe routing involves determining the most efficient path for pipes while considering ease of operation, maintenance, accessibility, and safety. This ensures minimal pressure drop, uniform flow distribution, and efficient use of space. Support design involves selecting appropriate pipe supports, hangers, and anchors to ensure the structural stability and integrity of the piping system.

Material Selection and Specifications: Piping and layout engineers are responsible for selecting appropriate pipe materials suitable for specific process conditions, including temperature, pressure, corrosion resistance, and environmental factors. Material specifications and standards, such as ASTM and ASME, are carefully considered to ensure compatibility and reliability.

Clash Detection and Interference Management: Advanced 3D modeling and design review software aid in clash detection and interference management. Piping and layout engineers work in collaboration with other disciplines to identify and resolve clashes between piping systems, equipment, and structural elements, ensuring smooth installation and operation.

Stress Analysis: Piping stress analysis is conducted to ensure that the piping systems can withstand operating conditions, pressures, and temperatures without failure or excessive deflection. Finite element analysis (FEA) techniques are employed to assess stress levels and optimize pipe supports, expansion joints, and flexibility requirements.

As-Built Documentation: Piping and layout engineering involves generating accurate as-built documentation, including piping isometrics, bills of materials, and piping and instrumentation drawings (P&IDs) as per the final installation. This documentation is essential for maintenance, future modifications, and compliance requirements

7. Safety and Environmental Engineering:

Safety and environmental engineers play a crucial role in ensuring the compliance of the facility with safety regulations and environmental standards. They collaborate with other engineering disciplines to identify potential hazards, develop safety protocols, and incorporate safety features into the design. Safety and environmental engineers also assess the environmental impact of the facility and work on minimizing emissions, waste generation, and environmental risks.

8. Project Engineering Management:

While not strictly an engineering discipline, project management is instrumental in the coordination and integration of the various engineering disciplines involved in detailed design engineering. Project managers ensure effective communication, manage timelines, budgets, and resources, and oversee the overall progress of the project. They facilitate the interfaces between different disciplines, ensuring that the design is comprehensive, coherent, and meets client expectations.

(There are still a few more engineering disciplines, specific to certain markets, which are not included)

In summary, detailed design engineering in the oil and gas sector involves the collaboration and interface of various engineering disciplines, including process, mechanical, civil and structural, electrical, instrumentation and control, piping and layout, safety and environmental, and project management. Together, these disciplines work towards developing a holistic and optimized design that ensures the safe, efficient, and environmentally compliant operation of oil and gas facilities.

Engineering disciplines in addition support the procurement process during the execution.

Engineering disciplines prepare the material requisitions and perform the technical bid evaluations during the procurement.

Post-purchase order placement, engineering disciplines perform the vendor document review process.

The involvement of engineering even extends during the construction phase and commissioning phase.

Thus, engineering becomes a fulcrum of EPC execution.

Piping and other supports are usually designed together to optimize. In some cases, the piping and E&I supports are designed separately depending on the size of the project.

In a multi-center execution scenario, the procurement is generally centralized, to reduce cost and increase quality.

Certain critical analyses like vibration analysis, and ergonomics engineering, are sometimes outsourced.

> **TITBITS**
>
> DATA is a raw content, usually free from any context.
>
> INFORMATION is data that has been placed in proper context, often with a comparative analysis.
>
> KNOWLEDGE is when the information is further distilled to become more insightful or actionable.
>
> WISDOM is the assimilation of knowledge and insights into a powerful combination of intelligence.

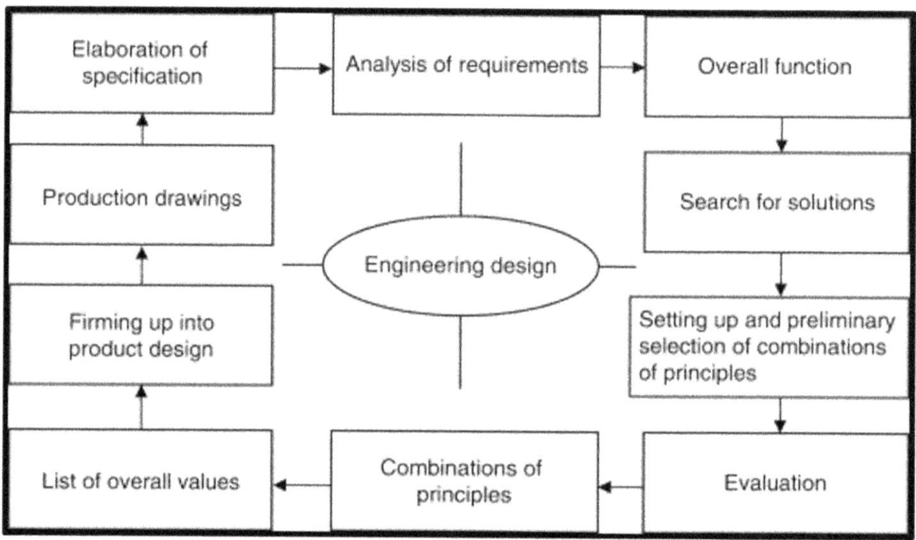

Importance of design engineering

Detailed design engineering plays a pivotal role in the oil and gas sector, ensuring the efficient and effective functioning of every aspect of the industry. The aim is to highlight the importance of detailed design engineering, its role in optimizing operations, and its impact on safety, productivity, and environmental sustainability.

1. Designing Facilities and Infrastructures:

In the oil and gas sector, detailed design engineering covers the planning, designing, and development of critical facilities and infrastructures such as refineries, drilling platforms, storage tanks, pipelines, and offshore production systems. These designs must guarantee safe operations, efficient resource management, and compliance with industry standards and regulations.

2. Safety Considerations:

Detailed design engineering focuses on incorporating safety features and measures within all oil and gas operations. This includes analyzing potential hazards, implementing robust safety protocols, and ensuring the resilience of infrastructure to withstand extreme conditions. By considering safety at the design stage, risks are mitigated, minimizing accidents, and protecting both human lives and the environment.

3. Efficiency and Productivity Enhancement:

Efficient design engineering aims to optimize oil and gas production processes, reducing operational costs and maximizing productivity. This involves incorporating automation technologies, streamlined workflows, and intelligent systems to minimize downtime, enhance equipment reliability, and improve overall efficiency. Proper design engineering allows for the smooth operation and maintenance of equipment, resulting in increased production output and reduced operational losses.

4. Environmental and Sustainability Considerations:

Detailed design engineering plays a crucial role in ensuring sustainability in the oil and gas sector. By integrating environmentally friendly practices and utilizing efficient technologies, it becomes possible to minimize carbon emissions, reduce waste generation, and promote resource conservation. Design engineers focus on incorporating renewable energy solutions, implementing waste management strategies, and constructing sustainable infrastructure, aligning the sector with global environmental goals.

5. Mitigating Risks and Challenges:

The oil and gas industry faces various risks such as geotechnical conditions, offshore drilling hazards, and extreme weather events. Detailed design engineering anticipates and addresses these challenges during the planning phase. Thorough assessments, including geological surveys and risk analysis, help engineers devise appropriate strategies to mitigate potential risks and ensure the longevity and success of the projects.

Detailed design engineering holds the key to the success and sustainability of the oil and gas sector. By focusing on safety, efficiency, productivity, and sustainability, design engineers contribute to the smooth functioning and growth of the industry. Through their expertise and meticulous planning, potential risks are mitigated, operations are optimized, and environmental impacts are minimized. As the oil and gas sector continues to evolve, detailed design engineering remains essential for continued progress and securing a sustainable future.

Engineering software

During detailed engineering in the oil and gas sector, various engineering software tools are utilized by different disciplines to enhance efficiency, accuracy, and collaboration. Here is a list of commonly used engineering software tools, discipline-wise, along with their purposes:

1. Process Engineering:

- Aspen HYSYS: A process simulation software used for modeling and optimizing process units and systems.
- PRO II: Another process simulation software that provides comprehensive modeling, optimization, and analysis of process units and processes.

- PipeFlo: Software for hydraulic calculations and network analysis of piping systems.

2. Mechanical Engineering:

- AutoCAD: A computer-aided design (CAD) software used for creating and modifying 2D and 3D designs, including equipment layouts and piping arrangements.
- Caesar II: Software for stress analysis of piping systems, ensuring compliance with piping codes and standards.
- PV Elite: Used for the design and analysis of pressure vessels, heat exchangers, and storage tanks.
- SolidWorks: CAD software utilized for mechanical design, modeling, and rendering of equipment and machinery.

3. Civil and Structural Engineering:

- STAAD.Pro: A structural analysis and design software that allows for the modeling, analysis, and design of various structural elements and systems.
- AutoCAD Civil 3D: Software for civil engineering design, including road and site design, land development, and terrain modeling.
- Tekla Structures: A 3D modeling software used for detailed structural analysis and design, particularly for steel and concrete structures.

4. Electrical Engineering:

- AutoCAD Electrical: A specialized version of AutoCAD for electrical design, providing tools for creating electrical schematics, panel layouts, and control diagrams.
- ETAP: Software used for electrical power systems analysis, including load flow, short-circuit analysis, arc flash, and relay coordination.

- EPLAN Electric P8: An electrical engineering design software that facilitates the creation of electrical schematics, panel layouts, and control cabinet designs.

5. Instrumentation and Control Engineering:

- AVEVA Instrumentation: Software for the detailed design, configuration, and documentation of instrumentation, control systems, and safety instrumented systems.
- Honeywell UniSim Design: Process simulation software with built-in control system design capabilities, enabling engineers to perform integrated process and control system design.

6. Safety and Environmental Engineering:

- PHAST: Software for hazard identification and risk assessment, used for analyzing and mitigating potential safety hazards.
- COMSOL Multiphysics: A simulation software used for modeling and analyzing complex physical systems, including environmental impact assessments and fluid dynamics simulations.

Engineering Software Validation

Engineering software validation refers to the process of assessing and confirming the accuracy, reliability, and suitability of engineering software applications for their intended purpose. It involves rigorous testing, analysis, and verification to ensure that the software performs as expected, meets design specifications, and produces reliable results.

During software validation, engineers or validation specialists conduct a series of tests and evaluations to assess various aspects of the software, including its functionality, performance, user interface, and adherence to industry standards or regulatory requirements. The validation process may involve both functional (does the software perform its intended tasks?) and non-functional (how well does the software handle large datasets or high traffic?) testing.

The main objectives of engineering software validation are to:

1. Ensure Accuracy: Validate that the software's calculations, simulations, or analytical capabilities produce results that are consistent, precise, and per established engineering principles, mathematical models, or algorithms.
2. Assess Reliability: Verify that the software operates reliably under different scenarios, handles various inputs and conditions without errors or crashes, and provides consistent output across repeated executions.
3. Confirm Compliance: Ensure that the software adheres to relevant industry standards, regulatory requirements, or best practices. This includes assessing the software's documentation, data management, security measures, and user interfaces for compliance.
4. Validate Usability: Evaluate the software's user interface, functionality, ease of use, and overall user experience to ensure that it is intuitive, efficient, and accessible to the intended users.
5. Identify and Resolve Deficiencies: Through rigorous testing, validation helps identify and address any bugs, glitches, errors, or deficiencies in the software. Feedback from users, stakeholders, and industry experts may also contribute to software improvements and enhancements.

Engineering software validation is essential to ensure that engineering professionals can rely on software tools for accurate analysis, design, simulation, and decision-making. It provides confidence in using the software to support engineering activities, reduces risks associated with inaccuracies or faulty outcomes, and enhances the overall efficiency and effectiveness of engineering processes.

In the case of in-house developed software (for example, Excel spreadsheets) Validation details shall be maintained at the department level.

Detailed engineering design

In general, detailed engineering design includes the following stages:

Checking specifications.

Defining subsystems, checking, and assembling parts.

Finalizing individual parts and accomplishing technical calculations.

Checking conformity with the standards.

Preparing and reviewing documents.

Concurrent Engineering

A method of designing and developing products, also known as simultaneous engineering, in which the different stages run simultaneously, rather than consecutively.

Optimization in Engineering Design

Optimization is often used during the engineering design process. It is a systematic process that uses design constraints and criteria to find an optimal solution. A wide range of optimization techniques and methods is available for researchers and designers, and they are selected per the nature of the optimization tasks, applications involved, and the designer's expertise. Therefore, despite the importance of optimization for engineering design, the introduction of various optimization techniques is not within the scope of this website, and you can find multiple resources for the optimization methods elsewhere.

Optimization in project execution

Engineering optimization plays a critical role in the Engineering, Procurement, and Construction (EPC) environment within the oil and gas sector. It involves applying optimization techniques to enhance the design, operations, and project management aspects, ultimately leading to improved efficiency, cost-effectiveness, and overall project success.

In the oil and gas sector, engineering optimization focuses on achieving key objectives, such as maximizing production output, minimizing operational costs, ensuring safety and environmental compliance, and meeting project deadlines. It navigates complex challenges, including system integration, resource allocation, risk management, and quality control.

Within an EPC environment, engineering optimization in the oil and gas sector encompasses various areas:

1. Facility Design and Layout: Optimization techniques help engineers design plants, refineries, pipelines, and offshore platforms while considering factors like process flow, equipment placement, and accessibility. This leads to efficient use of space, reduced construction costs, improved maintenance access, and enhanced overall operability.

2. Process Optimization: Engineers strive to optimize process parameters, such as temperature, pressure, and flow rates, to maximize production yield, energy efficiency, and product quality. Advanced modeling and simulation tools aid in identifying optimal operating conditions and process configurations to minimize resource consumption and emissions.

3. Supply Chain Optimization: Optimization methods assist in optimizing the flow and logistics of materials, equipment, and manpower across the entire supply chain. This enables efficient procurement, inventory management, transportation planning, and resource allocation, reducing costs and ensuring timely project completion.

4. Project Planning and Scheduling: Optimization techniques are utilized to develop realistic schedules and resource allocations, considering tasks, dependencies, resource availability, and project milestones. By optimizing the project schedule, engineers can

minimize overall project duration, reduce downtime, and enhance productivity.

5. Risk Management: Engineering optimization incorporates risk analysis and mitigation strategies to identify potential bottlenecks, vulnerabilities, and uncertainties in the project. This facilitates proactive planning and decision-making to address risks, ensuring project safety and success.

By leveraging engineering optimization in an EPC environment within the oil and gas sector, project stakeholders can achieve improved operational efficiency, reduced costs, enhanced safety and compliance, and increased project profitability. It enables informed decision-making, facilitates better resource management, reduces project delays, and ensures that engineering designs and operations are optimized to the fullest extent.

Value Engineering

Value engineering, also known as value management or value analysis, is a systematic and collaborative approach used to identify and improve the value of a project, product, or process. It focuses on maximizing the desired benefits while minimizing costs and optimizing performance.

The goal of value engineering is to identify innovative solutions that maintain or enhance the functionality, quality, and reliability of a system while reducing overall costs. It involves a multidisciplinary team that analyzes various aspects, including design, materials, processes, and functionality, to find opportunities for improvement.

Value engineering follows a structured methodology that typically involves the following steps:

1. Planning: Define the project objectives, scope, and constraints. Establish clear criteria for evaluating value.
2. Information Gathering: Gather relevant data and document the current system's components, functions, and costs.

3. Analysis: Break down the system into its components and analyze their functions, performance, and costs. Identify any redundancy, inefficiencies, or opportunities for improvement.

4. Generation of Ideas: Encourage brainstorming and generate multiple ideas or alternatives for improving value. Consider factors such as performance, quality, safety, aesthetics, and cost.

5. Evaluation: Evaluate each alternative against the defined value criteria. Analyze their feasibility, benefits, and potential risks.

6. Development: Further develop and refine the selected alternatives. Consider potential impacts on schedule, performance, and other related factors.

7. Implementation: Implement the approved value engineering recommendations, ensuring proper communication and coordination among the stakeholders.

Value engineering aims to strike the balance between cost, performance, and functionality. By focusing on value rather than solely on cost reduction, it enhances the overall project outcome by optimizing resources, minimizing waste, and improving customer satisfaction.

Overall, value engineering is a powerful tool that enables organizations to extract maximum value from their projects, products, or processes while meeting the desired objectives and constraints. It fosters innovation, efficiency, and continuous improvement, ultimately leading to enhanced competitiveness and customer value.

Five mantras for better design:

1. Safety
2. Functionality
3. Constructability
4. Operability
5. Maintainability

Detailed Engineering development stages

The Engineering development has the following stages during project execution.

Design planning

Reading the contract documents and preparing execution plans is the start of the design planning stage.

The engineering execution plan is a key document prepared at this stage.

Design inputs

The FEED deliverables, Specifications, go by documents, Standards, and Lessons learned form the design input.

Design verification

(Also called as Checking process)

Design verification during detailed engineering is a crucial step in the engineering process to ensure that the design meets the intended requirements, specifications, codes, and standards. It involves a systematic review and analysis of the design to validate its accuracy, functionality, and compliance. The specific approach to design verification can vary depending on the industry, project complexity, and applicable regulations, but generally includes the following steps:

Self-check

Self-check is performed by the preparer and ensures that the requirements are fulfilled.

Single discipline check

The deliverable is then tossed to the discipline checker for a single discipline check. The competency of the checker is very key factor at this stage.

Multi-discipline check (or inter-discipline check)

Certain deliverables require multi-discipline checks. (For example, the P & ID, Plot plan, etc.). the deliverable is sent for various affected engineering disciplines and quality for performing the inter-discipline check.

The comments are then resolved by the preparer (or originator) and sent for client review.

Client Review

The client reviews the deliverable and makes comments if necessary.

if commented, the comments are resolved by the lead engineer and submitted again for approval.

once the deliverable is approved it is issued either for design or construction. (IFD or IFC)

Few deliverables may require review by a classification society or a third party, and this is complied with in the project.

Design verification process:

1. Review of Design Documents: The design team, typically consisting of engineers, architects, and specialists, carefully reviews all project documentation, including drawings, specifications, calculations, and technical reports. This comprehensive review helps to identify any inconsistencies, errors, or discrepancies that need to be addressed.
2. Cross-Checking and Calculations: The design team verifies that all calculations, equations, and formulas used in the design are accurate and in line with relevant standards and regulations. They cross-check critical design parameters, such as loads, dimensions, capacities, and configurations, to ensure consistency and reliability.
3. Simulation and Modeling: Computer-aided design (CAD) tools and software are widely used to create virtual models and simulate

the behavior of the design in different conditions and scenarios. This allows engineers to evaluate performance, analyze structural integrity, assess fluid dynamics, and optimize the design for efficiency and safety.

4. Prototype Testing: In some cases, physical prototypes or scaled models are built to perform specific tests and experiments. These tests validate the design's performance, endurance, and functionality under real-world operating conditions. Results from prototype testing are compared to design specifications to ensure the design's adequacy.

5. Peer Review and Independent Assessment: A review by experienced professionals who were not directly involved in the detailed design process is essential. Independent experts offer a fresh perspective, provide valuable insights, and identify potential flaws or areas of improvement that may have been overlooked during the design process.

6. Compliance with Standards and Regulations: Design verification also entails confirming that the design complies with all applicable standards, codes, regulations, and industry practices specific to the project and the jurisdiction in which it is executed. This includes safety regulations, environmental requirements, structural codes, electrical codes, etc.

7. Documenting Verification Results: All findings, assessments, and modifications resulting from the design verification process are documented. The verification report summarizes the verification activities performed, highlights any deviations, provides resolutions, and confirms the design's compliance with relevant requirements.

Design verification is an iterative process, and feedback from various stakeholders, including clients, regulatory authorities, and external experts, is often considered to ensure thorough examination and validation. By performing robust design verification during detailed engineering, potential errors, risks, and design shortcomings can be identified and rectified at an early stage, minimizing the potential for costly revisions or compromised project performance.

Design review

There are a variety of reviews that happen during the detailed engineering stage.

To highlight a few:

- 3D model review
- P & ID review
- Plot plan review
- HAZOP
- HAZID
- SIL
- Constructability

Generally, these reviews are performed by subject matter experts in the presence of the entire engineering team. This makes the design more robust and safer.

During the detailed design engineering phase of an oil and gas project, several design reviews are conducted to ensure the accuracy, completeness, and compliance of the design. The specific design reviews may vary depending on the project and company, but here are some commonly performed design reviews in the oil and gas sector:

1. Conceptual Design Review: This review occurs early in the design process and focuses on assessing the overall concept of the project. It ensures that the design meets the project's objectives, scope, and requirements. Key aspects that are examined include feasibility, cost-effectiveness, and operability.

2. Process Design Review: This review examines the process engineering design to ensure the safe and efficient operation of the plant or facility. It assesses factors such as process flow diagrams

(PFDs), heat and material balances, equipment sizing, equipment selection, and process safety considerations.

3. Piping and Instrumentation Diagram (P&ID) Review: P&IDs are detailed diagrams that depict the interconnections between equipment, instruments, and piping in the system. This review focuses on verifying the accuracy and completeness of the P&IDs, ensuring all equipment and instrumentation are correctly represented, and confirming compliance with codes and standards.

4. Mechanical Design Review: This review evaluates the mechanical engineering design of equipment, structures, and systems. It verifies that the design considers factors such as load requirements, material specifications, safety factors, corrosion protection, and appropriate standards and codes.

5. Electrical and Instrumentation Design Review: This review assesses the electrical and instrumentation engineering design of the project. It examines elements such as electrical equipment selection, cable routing, instrument specifications, control systems, safety instrumented systems, and compliance with relevant codes and standards.

6. Civil and Structural Design Review: This review focuses on the civil and structural engineering design of the project. It ensures that the design meets the project's requirements for foundations, buildings, pipe racks, supports, and other civil and structural components. Structural calculations, soil analysis, and compliance with design codes and standards are assessed.

7. Safety Design Review: Safety design reviews focus on evaluating the overall safety measures and features incorporated into the design. This includes assessing safety systems, emergency response plans, fire protection systems, hazard identification, and risk mitigation strategies.

8. HAZOP (Hazard and Operability) Review: HAZOP is a systematic technique used to identify and evaluate potential hazards and operational issues in the design. This review involves

a multidisciplinary team analyzing the design through a series of brainstorming sessions to identify deviations from normal operations that may lead to accidents or non-compliance with safety standards.

9. Constructability Review: This review focuses on assessing the design for ease of construction and installation. It involves collaboration between the design team and the construction team to identify potential constructability issues, optimize construction sequences, and improve project efficiency.

10. Design Interface Review: This review is conducted to ensure the integration and compatibility of various design disciplines and interfaces. It ensures that different aspects of the design come together seamlessly, such as electrical and mechanical interfaces, structural and piping interfaces, and interfaces between different systems.

11. 3D Model Review: A 3D model review involves assessing the digital representation of the project using three-dimensional modeling software. This review verifies the accuracy and integrity of the model by checking for clashes, interferences, and inconsistencies between different disciplines, such as piping, equipment, structures, and electrical systems. It ensures that the design is clash-free, accurately represents the layout, and allows for proper installation, maintenance, and operation.

12. Safety Integrity Level (SIL) Review: SIL review is conducted to ensure that the design and implementation of safety instrumented systems (SIS) meet the required safety integrity levels. It involves assessing the reliability and performance of SIS functions to mitigate potential hazards and risks. The review includes the following steps:

    - Identification of Safety Instrumented Functions (SIF): Review the process hazards and identify the critical safety functions that require SIS to mitigate those hazards.

- Determination of Required SIL: Analyze the risks associated with each SIF and determine the safety integrity level required to achieve an acceptable risk reduction.
- Verification of SIS Design: Evaluate the design elements of the SIS, including sensor selection, logic solvers, final elements, and voting arrangements. Verify that the design complies with relevant standards and meets the required SIL.
- Calculation of PFD/PFH: Perform calculations to determine the Probability of Failure on Demand (PFD) or Probability of Failure per Hour (PFH) for each SIF. This quantifies the reliability and performance of the SIS and ensures it meets the targeted SIL.
- Documentation and Reporting: Document the findings of the SIL review, including the identified SIFs, SIL target, design verification, and PFD/PFH calculations. Report any deviations, recommendations, or actions needed to achieve the desired SIL.

13. HAZID (Hazard Identification) Review: HAZID is a systematic review that aims to identify potential hazards and associated consequences in a facility, process, or system. The review focuses on understanding the possible causes and consequences of hazardous events. The HAZID review typically involves the following steps:

- Review of Design Basis: Familiarize with the project's design basis, including process flow diagrams, operational procedures, and hazard identification and assessment criteria.
- Identification of Hazards: Use a combination of brainstorming and structured analysis techniques to identify potential hazards related to the project, such as fire, explosion, toxic release, environmental impact, or equipment failure.
- Assessment of Consequences: Evaluate the potential consequences of identified hazards, considering factors

such as personnel safety, environment, asset integrity, and reputation. This assessment provides insights into the severity of each hazard.

- Documentation of Risks: Document the identified hazards, associated risks, and their potential consequences. This information helps in formulating risk mitigation strategies and designing safety measures.
- Risk Evaluation: Assess the identified risks based on their severity, likelihood, and detectability. Prioritize the risks for further analysis, control measures, and mitigation actions.
- Reporting and Recommendations: Prepare a HAZID report summarizing the findings, risks, and recommendations. The report ensures that the identified hazards are properly addressed, and appropriate actions are taken to reduce risks to an acceptable level.

Performing SIL and HAZID reviews help ensure the safety and integrity of oil and gas projects by systematically identifying hazards, assessing risks, and implementing necessary control measures.

Design Validation

Generally, the design reviews are considered as design validation. This shall be clearly defined in the engineering execution plan.

1. Equipment and System Design Validation:

- Validate the design of equipment, such as pumps, compressors, heat exchangers, and vessels, ensuring they meet project specifications, industry codes, and client requirements.
- Verify the compatibility and functionality of interconnected systems, such as piping, instrumentation, electrical, and control systems, through thorough design checks and simulations.

2. Piping and Instrumentation Diagram (P&ID) Validation:

- Validate the accuracy and completeness of P&IDs, ensuring they accurately represent the desired process flow, equipment, and control instrumentation.
- Review P&IDs to identify potential design issues or inconsistencies, addressing them before construction commences.

3. Structural Design Validation:

- Validate the structural integrity and stability of key components, such as platforms, supports, and foundations, ensuring they meet applicable codes and project-specific requirements.
- Perform structural analysis and calculations, verifying factors such as load-bearing capacity, vibration resistance, and safety margins.

4. Electrical System Design Validation:

- Validate the design and performance of electrical systems, including power distribution, lighting, grounding, and cable routing.
- Verify compliance with electrical codes, standards, and project specifications, ensuring the design can handle the required electrical loads.

5. Safety System Design Validation:

- Validate the design and functionality of safety systems, such as fire and gas detection systems, emergency shutdown systems, and safety instrumented systems.
- Verify that safety systems comply with regulatory requirements, industry standards, and project-specific safety guidelines.

6. Technical Documentation Validation:

- Validate the accuracy and completeness of technical documentation, including engineering drawings, specifications, datasheets, and calculations.
- Review documentation for consistency, correctness, and adherence to applicable standards and project requirements.

7. Design Interface Validation:

- Validate the interface compatibility and integration between different engineering disciplines, ensuring seamless coordination and functionality across disciplines.
- Ensure proper alignment between piping, instrumentation, electrical, and structural drawings to avoid clashes or conflicts during construction and installation.

By specifically focusing on design validation for engineering deliverables, oil and gas EPC projects can ensure that the various engineering components and systems meet the necessary requirements, functional specifications, and safety standards. This validation process helps to minimize design errors, optimize constructability, and deliver reliable and compliant engineering deliverables.

Design changes

Design change management plays a crucial role during detailed engineering execution. It is primarily the responsibility of the respective lead engineer and coordinated by the project engineer.

A design log is maintained by the project and monitored regularly for action items, and close out.

Certain changes may require the Client's approval and could impact either the COST or the SCHEDULE of the project.

Project design change management in the EPC (Engineering, Procurement, and Construction) sector of the oil and gas industry involves effectively managing changes and modifications to the project design. It encompasses processes and procedures to handle design variations, updates, or revisions that may occur during the project lifecycle.

1. Design Change Identification: Design change management begins with the identification of potential changes or modifications that may impact the project design. This includes changes in project requirements, changes in design specifications, value engineering opportunities, client requests, or unforeseen issues that may arise during project execution.

2. Impact Assessment: Identified design changes are then assessed for their potential impact on the project. This includes evaluating the implications of design changes on project scope, engineering, procurement, construction activities, schedule, cost, quality, and other project parameters.

3. Change Request and Evaluation: Based on the impact assessment, design change requests are formally made, documented, and evaluated. Change requests should include a clear description of the proposed design change, rationale, potential benefits or drawbacks, and the resources required for implementation.

4. Change Approval and Authorization: Design change requests are reviewed and assessed by project stakeholders, including the project manager, engineering team, client, and other relevant parties. Approval and authorization processes are followed to determine whether the design change should be approved, rejected, or deferred for further evaluation.

5. Design Change Implementation and Communication: Once a design change is approved, it is incorporated into the project documentation and communicated to the engineering team,

suppliers, subcontractors, and other relevant stakeholders. The necessary adjustments are made to the design drawings, engineering documents, specifications, and other project elements affected by the approved design change.

6. Change Tracking and Documentation: Design changes are systematically tracked and documented throughout the project lifecycle. This includes maintaining a change register, changelog, or change management system to record details of the design change, its impact, approval status, implementation date, and any associated actions or decisions.

7. Change Control and Monitoring: Design change control procedures are established to monitor and control design changes effectively. This involves ongoing tracking of approved design changes, assessing their impact on project progress and performance, managing related risks, and ensuring adherence to change management processes and quality management standards.

8. Lessons Learned and Continuous Improvement: Project design change management incorporates capturing and applying lessons learned from design change management activities. Feedback from past projects, design change experiences, and performance evaluations are utilized to enhance design change management processes, controls, and stakeholder engagement for future projects.

By efficiently managing design changes in the EPC oil and gas sector, organizations can respond to evolving project requirements, improve project outcomes, mitigate risks, and reduce delays and cost overruns. Effective design change management helps ensure project success, alignment with client expectations, and the delivery of high-quality assets and facilities.

List Engineering deliverables Basic, FEED, and Detailed engineering

Engineering deliverables at BASIC, FEED and Detail Engineering stages

Discipline	Deliverable/activity	BASIC	FEED	DETAIL
General	Basic Engineering Design Data		✓	
	Procurement Plan with identification of LLI		✓	
	Project Time schedule	preliminary	update	final
	Project Equipment list		With dimensions & weights	
Safety & Environment	HAZOP report		✓	✓
	HAZOP close-out report (see note)		✓	✓
	HAZID		✓	
	HAZID close-out report		✓	
	Quantitative Risk Assessment (QRA)		for permitting and blast design	✓
	Hazardous area classification drawings		IFD	✓
	Fire water demand calculation		prelimnary	final
	Fire & Gas detection layout			✓
	Fire Water pumps and network P&IDs		IFD	IFC
	Fire water network layout drawing		IFD	IFC
	Safety concept & philosophies		✓	
	Environment requirements specification		✓	
	Environment impact Assessment (EIA)		for permitting	✓
Process	Process Design basis	✓		
	Process Design criteria	✓		
	Process Flow Diagrams (PFDs)	✓	✓	✓
	Heat & Material Balances (HMB)	✓	✓	✓
	Process Equipment list	Sized except HX	Sized	✓
	P&IDs	Process	Process + Utility	Process + Utility + Packages
	P&ID review	✓	✓	✓
	Process description		✓	
	Operating manual			✓
	Equipment Process Data Sheets	Process Equipment only	Process + Utility Equipment, flare, drain, etc.	
	Instrument Process Data Sheets		Control & on/off valves, analysers, flowmeters	all
	Packaged units duty specifications	Main	all	
	Emergency shutdown philosophy		✓	
	Causes & Effects diagrams		✓	
	Heat exchangers thermal design & data sheets		✓	
	Utility consumption	preliminary	update	final
	Flare relief load	preliminary	update	final
Equipment/ Mechanical	Equipment specifications		✓	
	Equipment mechanical design		✓	
	Equipment Mechanical Data Sheet		✓	
	Material requisition for inquiry		For main equipment	For all
	Technical Bid Tabulation		For main equipment	For all
	Material requisition for purchase		For LLI	For all
	Vendor drawings			✓
Plant Layout	Unit Plot Plan	typical	IFD	IFC
	3D modelling		Equipment + main process lines only	✓
Piping Installation	Piping routing drawings		✓	
	Piping studies and Layout drawings			✓
	Piping General Arrangement drawings			✓
	Isometric drawings			✓
Piping/Material	General specification for piping materials		✓	
	Piping classes summary		✓	
	Piping Material Classes Specifications		✓	
	Material selection diagrams	✓		
	Special piping items specification			✓
	Standard drawings		design	installation
	Piping MTO		from P&ID and Plot Plan	from Piping layout studies/3D model

Discipline	Deliverable/activity	BASIC	FEED	DETAIL
Piping/Stress	Design specification (piping stress design basis)		✓	
	Piping stress calculations		Simplified calculation of critical lines with impact on Equipment layout	✓
	Piping support drawings and list			✓
Instrumentation & Control	Instrumentation & automation design specification		✓	
	Systems specifications		✓	
	Instrument data sheets		ON/OFF valves, control valves, flowmeters, analysers	all
	Instrument specification for packaged units		✓	
	System architectural drawing		✓	
	Systems I/O sizing		✓	
	Instrument list		✓	
	Material requisitions for Systems		For inquiry	For purchase
	Material Take-Off			✓
	Cable schedule			✓
	Loop diagrams			✓
	Standard drawings		✓	
	Control and technical rooms Equipment Arrangement drawings		Preliminary (size of building only)	✓
	Cable routing drawings		Main routings and JB locations only	Main and secondary routing
Civil	Soil investigation specification		✓	for additional soil investigations
	Review soil investigation report		preliminary report	final report
	Earthworks drawings		✓	
	Underground networks drawings		General (1:200 scale)	Area (1:50 scale)
	Civil & Structural design basis & criteria		✓	
	Specifications for civil works		Main	all
	Guide /outline drawings		if needed to perform MTO	
	Design drawings			✓
	Equipment foundation drawings		Typical	✓
	Calculations & calculation notes			✓
	Drainage network calculation		✓	
	Concrete/steel standard drawings		design	construction
	Buildings architectural drawings		✓	
	Material take-off		preliminary	up-date
Electrical	Electrical design criteria		✓	
	Single Line Diagram		General	Switchgear's
	Electrical consumers list & power balance		preliminary	final
	Equipment general specification		✓	
	Electrical specification for packaged units		✓	
	Equipment data sheet		HV, MV switchgear only	all
	Material requisitions for Electrical Equipment		For inquiry	For purchase
	Standard drawings		design	installation
	Specification for bulk			✓
	MTO		preliminary (for cables only)	all
	Cable schedule		preliminary	final
	Substation Equipment arrangement drawings		prelimnary (for sizing of sub-station only)	
	Cable routing drawings		Main routings only	all routings
	Calculations		Load summary	all
Painting, Coating, Insulation	Specification		✓	
	Standard drawings			✓
Cost estimate	+/-30-40% cost estimate	✓		
	+/-15-20% cost estimate		✓	

International standards

In the oil and gas sector, certain API (American Petroleum Institute) and ASME (American Society of Mechanical Engineers) are used widely for design. Here are some key API and ASME certifications and stamps relevant to the oil and gas sector:

1. API Spec Q1: This certification establishes the quality management system requirements for organizations manufacturing products or providing services in the oil and gas industry. It demonstrates that a fabrication shop has implemented robust quality control measures to meet industry standards.

2. API Spec 2B: This specification covers the fabrication of structural steel pipe supports used in offshore platforms or other oil and gas structures. It ensures that the fabrication shop meets the required specifications for welding, dimension control, and material quality.

3. ASME Section VIII Division 1: This code applies to the fabrication of pressure vessels used in oil and gas operations. Fabrication shops must hold an ASME U stamp to show compliance with the code, indicating their ability to design, manufacture, and inspect pressure vessels according to ASME standards.

4. ASME B31.3: This code specifies the requirements for the design, fabrication, and inspection of process piping systems. Fabrication shops may require the ASME PP stamp (Process Piping) to certify their competence in fabricating process piping systems per the ASME B31.3 code.

5. API Spec 6A: This specification covers the requirements for the design, manufacturing, and testing of wellhead and Christmas tree equipment used in oil and gas exploration and production. Fabrication shops involved in the production of this equipment may need API 6A certification to validate compliance with the standard.

6. API Spec 5L: This specification covers the requirements for the fabrication of seamless and welded steel line pipe used in the transportation of oil, gas, and other fluids. Fabrication shops should ensure compliance with this standard to meet the specifications for line pipe fabrication.

The list is long considering, ASTM, AWS, ISO, DIN, DNV, etc. Let me stop with the above.

It is important to note that specific projects may have additional requirements beyond the certifications and stamps mentioned above. Clients and regulatory bodies may impose additional codes, standards, or specifications based on project-specific needs, environmental conditions, and equipment requirements. Therefore, fabrication shops need to evaluate project requirements and adhere to relevant codes, standards, and specifications to ensure quality, safety, and compliance in the oil and gas sector.

TITBITS

Ganbatte, often translated as "do your best" or "keep going," encompasses the spirit of perseverance, determination, and resilience that is deeply ingrained in Japanese culture. It reflects the belief that through sheer effort and dedication, one can overcome challenges, achieve success, and constantly strive for improvement.

The concept of ganbatte emphasizes the importance of not giving up, even in the face of adversity or setbacks. It encourages individuals to give their best effort, to work diligently towards their goals, and to maintain a positive attitude. It is also a word of encouragement used to inspire and support others in their endeavors. By embodying ganbatte, individuals can unlock their full potential and accomplish remarkable achievements.

Embrace resilience: Cultivate a mindset that accepts challenges as opportunities for growth and learning.

- Set goals: Define clear and achievable goals to provide direction and focus.
- Maintain a positive attitude: Adopt a mindset of optimism and determination, even in the face of obstacles.
- Support and uplift others: Encourage and motivate others in their pursuits, fostering a sense of community and collaboration.

The role of licensors

Here are some key aspects of the licensor's role in the oil and gas industry in the EPC market:

1. Proprietary Technology: Licensors are often owners of valuable and proprietary technologies used in oil and gas projects. These

technologies can include refining processes, petrochemical manufacturing processes, liquefaction technologies, or advanced extraction methods. Licensors grant licenses to EPC companies or clients to use their exclusive technology for a specific project.

2. Engineering and Design Support: Licensors provide extensive engineering and design support to EPC companies during the project execution. They work closely with the EPC team to ensure the proper integration of their technology into the overall project design. This collaboration includes providing detailed technical specifications, design guidelines, and assistance in adapting the technology to meet project-specific requirements.

3. Technology Transfer: As part of their role, licensors transfer their proprietary technology to the EPC company. This transfer involves training the EPC team on the proper implementation and operation of the technology. The licensor will also provide documentation, manuals, and ongoing support to ensure the successful adoption of their technology.

4. Quality Control and Compliance: Licensor companies play a critical role in maintaining quality control and compliance with industry standards. They supervise the implementation of their technology, ensuring that the EPC company adheres to specific guidelines and specifications. This involvement helps guarantee that the project meets all necessary safety, environmental, and regulatory requirements.

5. Performance Guarantee: Licensors often provide performance guarantees for their technology. This means that they assure certain predefined performance parameters will be achieved when their technology is properly implemented. Performance guarantees provide confidence to the EPC company and the end client, assuring them of the technology's effectiveness and reliability.

6. Continuous Support: Licensors are generally available for continuous support during the project's lifecycle. They provide technical assistance, troubleshooting, and guidance throughout the construction, commissioning, and operation stages to address any issues that may arise related to their technology.

Overall, the licensor's role in the oil and gas industry in the EPC market ensures the successful integration of proprietary technology into projects. Their expertise, support, and guidance play a vital role in maximizing project efficiency, safety, and overall performance.

Structural Analysis

Structural Analysis for Offshore Platforms: Ensuring Safety and Resilience

Offshore platforms are engineering marvels, but their resilience and safety depend on rigorous structural analysis. Here's a brief overview of the key analyses involved:

1. Static & Dynamic Analysis: Assess stability, stress, and deformation under static and dynamic loads.
2. Seismic Analysis: Evaluate earthquake resistance, including vertical and horizontal forces.
3. Finite Element Analysis (FEA): Detailed stress and deformation analysis for precision.
4. Buckling & Fatigue Analysis: Prevent buckling and assess fatigue under cyclic loading.
5. Foundation Analysis: Evaluate seabed interaction, accounting for soil properties.
6. Risk & Reliability Analysis: Assess overall structural reliability considering uncertainties.

Towing, Transport, and Mooring:

7. Hydrodynamic Analysis: Predict platform response during towing and transport.

8. Mooring System Design: Analyze anchor points and dynamic responses for secure mooring.
9. Structural Health Monitoring: Real-time monitoring ensures early issue detection.

Hazards: Boat Impact, Ice Impact, and Fire:

10. Boat Impact Analysis: Assess vessel collision forces and recommend mitigation measures.
11. Ice Impact Analysis: Evaluate ice load impacts in icy environments.
12. Fire and Explosion Analysis: Analyze platform response to fire and blast loads for fire protection measures.
13. Collision Analysis with Other Offshore Structures: Examine potential collisions with neighboring structures.

These analyses are vital for the safety, resilience, and reliability of offshore platforms. They ensure these structures meet the demands of the energy industry and the challenges of marine environments. #StructuralEngineering #OffshorePlatforms #Safety #EngineeringAnalysis.

Technical HSE (also called Loss prevention engineering or Design Safety)

- Fire & Explosion Risk Analysis (FERA)
- Nonflammable Hazard Analysis (NFHA)
- Escape, Evacuation & rescue Analysis (EERA)
- Emergency system Survivability Analysis (EESA)
- Building Risk Assessment (BRA)
- Quantitative Risk Assessment (QRA)

HAZID (Hazard Identification)

HAZID is a technique for early identification of Occupational Health, Safety, and Environmental (HSE) related major accident events (MAEs)/ Hazards of a project.

QRA (Quantitative Risk Assessment)

QRA is a formal and systematic approach to estimating the likelihood and quantitative consequences of major accident events (MAEs) and expressing the results in terms of risk to personnel, assets, and the environment.

EERA (Escape, Evacuation & rescue Analysis)

EERA is done to ensure controlled evacuation to the identified Safe zone in situations of hazardous events for all personnel on board and to facilitate the rescue of distressed or injured personnel. EERA ensures timely evacuation and also calculates whether the time taken for Escape & Evacuation is adequate.

DOS (Dropped Object Study)

A dropped Object Study is done, to calculate the dropped object frequencies from the topsides, to assess dropped object impact energy for the laydown area during the transfer from/to supply vessels, to estimate impact energy for material handling activities within the platform, and to identify vulnerable areas exposed to the dropped object hazard.

This list is not comprehensive. In actual practice, depending on the project the studies may vary.

Don't get confused "Technical HSE" with "Site HSE". They are different.

Some organizations call this department "Loss prevention".

Packaged equipments

Packaged equipment plays a vital role in the oil and gas industry, serving as essential components of various processes and systems. Managing packaged equipment effectively within the EPC (Engineering, Procurement, and Construction) framework is crucial to ensure seamless project execution and operational success.

Packaged equipment refers to pre-engineered systems or modules that are fabricated, assembled, and tested in a controlled environment before being delivered to the project site. These equipment packages can include compressors, pumps, generators, separators, heaters, and other crucial components like piping and structure required for oil and gas operations.

The management of packaged equipment in the EPC process involves several key aspects:

1. Equipment Specification and Selection: During the project's engineering phase, the EPC team collaborates with the client to define the project requirements. This includes selecting the appropriate packaged equipment based on operational needs, technical specifications, industry standards, and safety considerations.

2. Procurement and Vendor Management: Once equipment requirements are identified, the EPC team manages the procurement process, including the selection and evaluation of reliable vendors or suppliers. This involves ensuring compliance with contractual requirements, tracking delivery schedules, and conducting quality inspections.

   – Equipment Specification and Evaluation: The EPC team works closely with the client to define the technical specifications and performance requirements for the packaged equipment. Clear specifications ensure that the selected equipment meets the project's needs and complies with industry standards, codes, and regulations.

   – Vendor Prequalification: The EPC team prequalifies potential vendors based on their experience, capabilities, financial stability, and quality management systems. This helps ensure that the selected vendors can deliver equipment that meets the project's quality and performance expectations.

- Request for Quotations (RFQ): The EPC team issues RFQs to prequalified vendors, clearly outlining the technical requirements, delivery timeframes, and commercial terms. Vendors are requested to submit detailed proposals that include equipment specifications, pricing, and any necessary supporting documentation.
- Bid Evaluation and Vendor Selection: The EPC team evaluates the received proposals based on technical compliance, commercial terms, delivery schedules, warranty provisions, and overall value for money. A thorough evaluation helps in selecting the most suitable vendor(s) for each specific equipment package.
- Negotiation and Contracting: Once the preferred vendor(s) are identified, the EPC team enters into negotiations to finalize the commercial terms, scope of supply, warranties, delivery schedules, inspection and testing requirements, and payment terms. The outcome of these negotiations is formalized in a contract that binds both the EPC contractor and the vendor.
- Quality Assurance and Inspection: Throughout the procurement process, the EPC team conducts regular quality assurance activities. These include vendor audits, technical document reviews, and inspection visits to ensure compliance with project specifications, codes, and standards. Inspection and testing plans are established to verify the quality and performance of the equipment during various stages of fabrication and assembly.
- Expediting and Delivery Management: The EPC team closely monitors the progress of equipment manufacturing and coordinates with vendors to ensure timely and successful delivery. Expediting activities may include production progress reviews, factory acceptance tests (FAT), and arranging third-party inspections when necessary.

    – Vendor Performance Management: The EPC team actively manages vendor performance throughout the procurement process. This includes tracking and reporting vendor progress, addressing any non-compliance, conducting performance reviews, and enforcing contractual obligations related to quality, delivery, and documentation.

The procurement process within packaged equipment management is critical to ensuring that the selected vendors meet the project's technical, commercial, and quality requirements. By effectively managing the procurement process, the EPC team can mitigate risks associated with equipment procurement, optimize costs, and ensure the successful delivery of packaged equipment for the oil and gas project.

3. Logistics and Transportation: Coordinating the logistics of transporting packaged equipment to the project site is critical. The EPC team plans and manages the transportation, including customs clearance, permits, packaging, and securing safe transportation methods to minimize the risk of damage during transit.

4. Installation and Integration: Proper installation and integration of packaged equipment is vital to ensure seamless functioning within the overall project system. The EPC team coordinates with installation contractors, ensuring adherence to design specifications, technical drawings, and safety protocols.

5. Testing and Commissioning: Before project handover, the EPC team oversees the rigorous testing, commissioning, and start-up of packaged equipment. This involves performance testing, functional checks, and verification of equipment compliance with project specifications and regulatory standards.

6. Maintenance and Repair: Throughout the project lifecycle, the EPC team establishes strategies for ongoing maintenance, operation, and potential repairs of the packaged equipment. This includes creating maintenance plans, spare parts management,

and implementing preventative maintenance practices to ensure equipment reliability and longevity.

By effectively managing packaged equipment within the EPC framework, oil and gas projects can benefit from streamlined operations, improved project scheduling, reduced risk of equipment failure, and enhanced overall project performance.

Valves

Valves – Valves are used to stop, open, or regulate the flow. There are many types of valves are used for different applications. Some of the common types are:

Gate Valve – This is the most common type used. It is usually manually operated and is designed for open or shut operation. It provides a minimum pressure drop of the fluid. Gate valves are most commonly used in drain services.

Ball Valve – used a ball with a hole in it which either lets flow pass through the hole in the ball or turns through 90 degrees so that no flow is possible. This type of valve is again used generally for open or shut operations where fluid leakage is not tolerated. Ball valves are most commonly used in process gas / liquid services.

Check Valve – It will allow flow to go one way but will not allow reverse. The common check valves are swing check in which a flapper lifts to permit flow and ball check which has a ball in it which lifts with flow. It is also called a non-return valve

Globe Valve – This type of valve is used for controlling or throttling flow. Globe valves are normally used in control valve by-pass lines.

Relief or Safety Valve – This is an automatic valve that opens when the system pressure exceeds design or operating parameters (set pressure). Without relief valves, plants and equipment could explode during periods of high pressure. The valve has a spring that holds the shut. The spring holds until a set pressure is obtained; and, when the pressure is more than the set pressure, the spring "gives" and allows the fluid to escape, thereby relieving the pressure. As the pressure reduces, the spring closes and shuts off the flow.

Control Valve – is usually an automatic valve that controls the flow in a piping system. A globe valve is generally used which opens, closes, or throttles on a signal from an instrument. Control valve's automatic operations normally by either pneumatically or hydraulically.

Butterfly Valve – This valve consists of a disc approximating in size to the bore of the adjacent pipeline. The disc is arranged to turn 90 degrees to control fluid flow and give a tight closure against a seat located in the valve body. This valve has a compact face-to-face dimension due to this principle of design. Butterfly valves are commonly used in fire water/ deluge systems and other utility services. Butterfly valves are not suitable for process lines.

Exotic materials

(Critical materials used)

In the EPC (Engineering, Procurement, and Construction) sector of the oil and gas industry, various exotic materials are employed to meet specific project requirements and mitigate challenges related to harsh operating conditions, corrosive environments, high temperatures, or unique properties. Here are some of the commonly used exotic materials in the oil and gas sector:

1. Duplex Stainless Steel (DSS): DSS offers excellent corrosion resistance, mechanical strength, and resistance to stress corrosion cracking. It is suitable for applications involving corrosive environments, high pressure, and high-temperature conditions.

2. Super Duplex Stainless Steel (SDSS): SDSS is a higher alloy version of duplex stainless steel, offering enhanced resistance to corrosion, erosion, and pitting. It is used in demanding environments such as offshore platforms, subsea pipelines, and chemical processing plants.

3. Inconel and Incoloy: These nickel-based alloys exhibit strong resistance to high temperatures, oxidation, and corrosion. They provide excellent performance in extreme environments, including high-temperature steam, acid gases, and sour oil and gas applications.

4. Titanium and Titanium Alloys: Known for their exceptional strength-to-weight ratio, titanium, and its alloys are employed in applications that require high strength, excellent corrosion resistance, and low weight. Titanium finds its utility in offshore platforms, heat exchangers, and components subjected to seawater exposure.

5. Hastelloy: Hastelloy alloys (such as C276, C22, and B2) are highly resistant to a wide range of corrosive and high-temperature environments, including sulfuric acid, hydrochloric acid, and seawater. They are used in chemical processing, offshore applications, and pollution control systems.

6. Monel: Monel alloys (e.g., Monel 400) offer excellent resistance to corrosion from various acids, alkaline solutions, and seawater. They are commonly used in offshore and marine applications, as well as in chemical industries.

7. Zirconium: Due to its excellent corrosion resistance, zirconium is used in highly corrosive environments containing strong acids, alkalis, or chlorides. It finds applications in chemical processing, nuclear power plants, and heat exchangers.

8. Nickel-based Clad Materials: Clad materials, such as corrosion-resistant alloys (CRA) lined with nickel-based materials (e.g., Inconel or Monel), are employed to provide superior corrosion resistance in equipment and vessels handling corrosive fluids or gases.

These are just a few examples of the exotic materials commonly used in the EPC oil and gas sector. The selection of material depends on project-specific requirements, process conditions, operating temperatures, and the presence of corrosive substances. Engineers, in collaboration with materials specialists, assess these factors to determine the most suitable material for each application within the project.

Basic details on some of the commonly used metals in the industry

Carbon Steel:

- Carbon steel contains primarily iron and carbon, with trace amounts of other elements such as manganese, phosphorus, sulfur, and silicon. The carbon content typically ranges from 0.05% to 2.0%.
- The mechanical properties of carbon steel can vary depending on the carbon content and heat treatment. Generally, carbon steel exhibits good strength, hardness, and toughness.

Stainless Steel:

- Stainless steel is an alloy predominantly composed of iron, chromium (min. 10.5%), and varying amounts of other elements like nickel, manganese, and molybdenum. Different grades of stainless steel may have specific alloy compositions.

- Chromium is the key element providing stainless steel its corrosion resistance. The addition of other elements enhances specific properties, such as nickel for increased resistance to acids and molybdenum for improved strength and pitting corrosion resistance.
- Stainless steel offers excellent mechanical properties, including high strength, hardness, and toughness, along with good ductility.

Monel:

- Monel is a nickel-based alloy consisting mainly of nickel (approx. 65-70%) and copper (approx. 20-30%) with smaller amounts of iron, manganese, and other elements.
- It provides exceptional resistance to corrosion, particularly in highly corrosive environments such as seawater. Monel alloys may have variations in composition but typically have high nickel content for increased corrosion resistance.
- Monel exhibits good mechanical properties, including high strength, hardness, and excellent resistance to extreme temperatures.

Inconel:

- Inconel is a family of nickel-based alloys, usually composed of nickel (min. 50%) and chromium, with additions of iron and other elements such as molybdenum, copper, and titanium depending on the specific grade.
- Inconel alloys are known for their exceptional resistance to high-temperature environments, oxidation, and corrosion. They retain their mechanical properties even at elevated temperatures.
- Inconel exhibits high strength, toughness, and outstanding mechanical properties at both room temperature and elevated temperatures.

Hastelloy:

- Hastelloy is a group of nickel-based alloys with varying compositions. They typically include nickel (min. 50%), chromium, molybdenum, and smaller amounts of other elements such as iron, cobalt, and tungsten.
- Hastelloy alloys are renowned for their exceptional corrosion resistance, especially in highly aggressive chemical environments.
- Hastelloy offers excellent mechanical properties, including high strength, hardness, and ductility.

Super Duplex Stainless Steel:

- Super duplex stainless steel is a high-performance alloy composed mainly of iron, chromium (approx. 24-26%), and molybdenum (approx. 3-4%), with additions of nitrogen and smaller amounts of other elements such as nickel and copper.
- The balanced composition of super duplex stainless steel provides superior corrosion resistance compared to standard duplex stainless steel, especially in highly corrosive environments.
- Super duplex stainless steel possesses excellent mechanical properties, including high strength, hardness, and good ductility.

These more detailed explanations provide insights into the chemical compositions and compositions of the mentioned materials. It's important to note that different grades and specifications within each category may exhibit variations in chemical compositions to meet specific requirements and industry standards.

Copper-nickel alloys, also known as cupronickel alloys, are a combination of copper and nickel with various amounts of other elements, such as iron and manganese, added to enhance specific properties. The most commonly used copper-nickel alloy is 90/10, which consists of 90% copper and 10% nickel.

In terms of chemical composition, copper-nickel alloys offer good corrosion resistance in various environments, particularly in seawater and marine atmospheres. The addition of nickel enhances the alloy's resistance to general and localized corrosion, as well as resistance to stress corrosion cracking. These alloys also possess excellent resistance to biofouling, making them an ideal choice for applications in marine environments.

Mechanically, copper-nickel alloys exhibit good strength, ductility, and toughness. They have high thermal conductivity and low electrical resistivity, making them suitable for heat exchangers, condensers, and cooling systems. These alloys also have excellent workability, allowing them to be easily formed, welded, and machined.

Due to their unique combination of properties, copper-nickel alloys find extensive use in various applications. They are commonly used in marine and offshore industries for shipbuilding, seawater desalination plants, and offshore oil and gas platforms. Additionally, they are utilized in heat exchangers, electrical components, and coinage.

Equipment criticality rating

Equipment criticality rating is performed basically for two purposes:

1. To arrive at the level of inspection required during the engineering and procurement phase.
2. To arrive at the O&M requirements during the plant operation stage.

Criticality Rating Considerations

The first consideration of the study is to identify the equipment, which would have a major negative impact on EHSS and operation during commissioning / after plant handover if flaws and defects escaped unnoticed during shop inspection and testing.

The second consideration of the study is to identify those equipment that are at greater risk of defects due to complexity in their fabrication and/or intricacy of design.

During criticality rating assessment, equipment shall be considered for the most stringent process design conditions to which it may be subjected.

Calculating the criticality rating:

Criticality rating shall be developed considering the cumulative risk the equipment/material is likely to exert on business during its lifetime due to four factors i.e., plant unavailability, EHS factors, Fabrication complexity and design Intricacy.

For each factor, the risk score could vary from 4 to 1 with 4 indicating very critical and 1 indicating least critical. The scores from all four factors are summed up to arrive at the overall criticality number for each equipment with a maximum overall tally of sixteen and a minimum of four.

The overall criticality score of each equipment is converted into an inspection level using the below table:

Level of Risk	Color Code/ Title	Overall Criticality Score Range	Inspection Level
Risk Level 1	Major	14-16	4
Risk Level 2	Significant	10 – 13	3
Risk Level 3	Minor	7- 9	2
Risk Level 4	Insignificant	4-6	1

Equipment criticality rating is performed to prioritize maintenance activities and allocate resources effectively. By assessing the criticality of equipment, organizations can identify the most important assets in their operations and focus their maintenance efforts accordingly. The process of equipment criticality rating typically involves the following steps:

1. Asset Identification: Identify the key equipment in the facility or system being assessed. This can include machinery, systems, and components that have a significant impact on production, safety, or operational efficiency.

2. Data Collection: Gather relevant information about each asset, such as its function, operating conditions, historical performance, maintenance history, and potential consequences of failure. This data can be obtained through equipment records, interviews with subject matter experts, and historical data analysis.

3. Risk Assessment: Evaluate the likelihood and potential consequences of failure for each asset. This can involve considering factors such as the asset's criticality to production, impact on safety, environmental risks, regulatory compliance, and financial implications.

4. Rating Criteria: Develop a set of criteria or a rating system to assess the criticality of each asset. This can involve assigning scores or rankings based on factors like safety implications, operational impact, financial considerations, customer satisfaction, and environmental concerns.

5. Rating Assignment: Apply the rating criteria to each asset and assign a criticality rating. This rating can be expressed using numerical scales, color-coded rankings, or other appropriate means to indicate the level of criticality.

6. Prioritization and Action Planning: Based on the criticality ratings, prioritize the maintenance activities and develop action plans for each asset. This can include determining the frequency and type of maintenance tasks, establishing preventive maintenance schedules, and allocating resources accordingly.

7. Review and Reassessment: Periodically review and reassess the criticality ratings as conditions change or new insights emerge. This allows organizations to adapt their maintenance strategies and priorities as needed to ensure optimal asset performance.

By performing equipment criticality ratings, organizations can identify and focus on the most critical assets, optimize maintenance efforts, improve reliability, reduce downtime, and enhance overall operational performance. It allows for a proactive approach to maintenance management by addressing the most critical risks to ensure the smooth functioning of the facility or system.

Asset Criticality Rating (ACA)

Asset criticality assessment in the oil and gas sector is a systematic evaluation of assets based on their importance to a company's operations and strategic objectives. It involves identifying and prioritizing assets according to their criticality or level of significance to the overall business.

The process of asset criticality assessment typically involves the following steps:

1. Asset Identification: Cataloging and documenting all relevant assets within the organization. This may include equipment, systems, infrastructure, facilities, and other components associated with the oil and gas operations.
2. Criticality Factors: Defining criteria or factors that determine the criticality of each asset. These factors may include safety implications, operational impact, regulatory compliance, environmental risks, financial implications, and production dependencies.
3. Criticality Rating: Assigning a rating or score to each asset based on the identified criticality factors. This rating serves as an indicator of the level of importance or impact that asset has on the overall operation.

4. Impact Assessment: Assessing the potential consequences of asset failures or disruptions. This involves considering the impact on safety, environmental compliance, production capacity, revenue generation, supply chain, workforce, and reputation.

5. Risk Mitigation: Developing strategies and plans to mitigate the risks associated with highly critical assets. This may include implementing preventive maintenance programs, redundancy measures, spare parts management, maintenance scheduling, and emergency response plans.

6. Decision-Making: Using the criticality assessment results to inform decision-making processes. This includes resource allocation, budgeting, capital expenditure prioritization, investment decisions, and determining maintenance and inspection priorities.

By conducting asset criticality assessments, oil and gas companies can optimize their asset management strategies, prioritize resource allocation, and ensure effective risk management. By understanding the criticality of each asset, organizations can focus their efforts and investments on maintaining and optimizing the most important components of their operations, thus enhancing safety, reliability, performance, and overall operational efficiency.

Categorization:

Tier 1

Failure has an immediate impact on or shutdown of multiple operations or systems. This failure will prevent capacity assurance due to operational, environmental, or quality issues. Equipment assigned this cursory criticality ranking (Rime Code) typically will have no redundancy, and identified issues must be addressed immediately to complete scheduled production targets and goals.

Tier 2

Failure results in limited production capabilities, or shutdown of, a single operation or system. Equipment assigned this ranking may have redundancy or established bypass equipment or systems but may limit capacity assurance. Although this equipment could become highly critical if the redundancy or bypass fails, identified issues should be planned and scheduled with a higher work order priority.

Tier 3

Failure results in an impact on, or shutdown of, a single operation or system. Equipment assigned to this ranking typically has redundancy or established bypass equipment or systems to complete the production schedule.

Tier 4

Failure has no immediate impact on capacity assurance. Some of these assets may have the maintenance strategy of Run-to-Failure associated with them. In contrast, others require addressing issues promptly through the normal planned workflow processes.

As built drawing

As-Built Drawings in EPC Projects: Ensuring Accuracy and Completeness in Engineering Documentation

In the realm of engineering within EPC (Engineering, Procurement, and Construction) projects, as-built drawings play a crucial role in documenting and capturing the final state of constructed facilities. These drawings represent a comprehensive record of the actual construction, modifications, and installations that have taken place during the project.

As-built drawings are created to accurately reflect the final condition of the project, highlighting any deviations or modifications from the original design or engineering drawings. They serve as a critical reference for future project operations, maintenance, modifications, and expansions. Here are key aspects related to as-built drawings:

1. Capturing Changes and Modifications:

As the construction progresses, modifications, adjustments, and changes are often made due to unforeseen circumstances or design enhancements. As-built drawings precisely document such changes, providing an accurate representation of the constructed facility.

2. Reflecting Actual Field Conditions:

As-built drawings are based on accurate measurements and field surveys to document as-built conditions. They consider the physical constraints and dimensional variations that may have occurred during construction, ensuring proper alignment with the real-world environment. Redline markups are normally sent from the fabrication yard to the engineering office, for updating.

3. Collaboration between Engineering and Construction Teams:

Creating reliable as-built drawings requires collaboration between engineering and construction teams. It is necessary to share critical information, document any changes made during construction, and ensure clear communication to accurately reflect the final state of the facility.

4. Ensuring Compliance and Safety:

As-built drawings are essential for verifying compliance with design specifications, building codes, regulations, and safety standards. They serve as a reference point for quality assurance and safety inspections, providing valuable documentation for future audits and assessments.

5. Supporting Operations and Maintenance:

Accurate as-built drawings become instrumental for operations and maintenance teams. They aid in identifying equipment, pipelines, routing, electrical systems, and other components, facilitating effective maintenance, troubleshooting, and asset management throughout the facility's lifecycle.

6. Providing a Basis for Future Development:

As-built drawings serve as a foundation for future expansions, modifications, or retrofitting projects. They provide essential information needed to plan and execute future activities, ensuring seamless integration and minimizing potential conflicts during subsequent project phases.

7. Maintaining Document Control:

Proper document control is vital for managing as-built drawings. Version control, storage, retrieval systems, and clear naming conventions are key factors to ensure effective document management and ease of access when required.

Accurate and comprehensive as-built drawings are essential for capturing and preserving the final state of constructed facilities in EPC projects. They provide a valuable reference for future engineering, operations, and maintenance activities, ensuring the facility's proper functioning, compliance, and adaptability as it progresses through its lifecycle.

TITBITS

Vision, mission, and value statements seem similar but have different audiences and purposes. Here are the differences:

- A vision statement is for internal company use A vision statement is a broad, aspirational goal that a company can work towards. It inspires employees and guides stakeholders.

- A mission statement is for internal and external use: A mission statement is a clear and concise summary of why a company exists. It's a reminder for staff and also provides the primary reason why clients or customers choose the company over competitors.

- A value statement is for external use: A value statement is an explicit list of the values for which a company stands. These are in clear correlation to the values of its customers or clients.

Companies that have clear and accurate vision, mission, and value statements are more likely to be successful. This is because everyone in the company knows what the overall goals are, which makes them easier to achieve. Clear company values allow people to work more efficiently and with greater clarity regarding their roles.

Statements by some of the famous companies:

Tesla: Accelerating the world's transition to sustainable energy

TED: Spread ideas, foster community, and create impact.

Microsoft: To empower every person and every organization on the planet to achieve more.

Google: Google's mission is to organize the world's information and make it universally accessible and useful.

Meta: Giving people the power to build community and bring the world closer together.

Lowe's: Together, deliver the right home improvement products, with the best service and value, across every channel and community we serve.

CHAPTER 6

Supply Chain Management

(Procurement, Expediting Logistics, and vendor management)

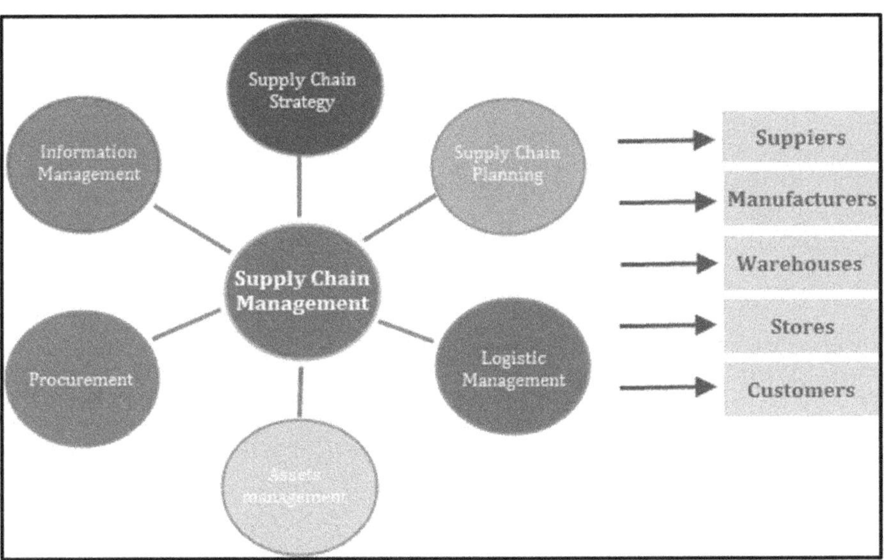

Procurement

Before getting into the detail, let us look at the difference between Procurement and Purchasing.

Procurement and purchasing are related terms that are often used interchangeably, but they have distinct meanings and functions within the overall supply chain process. Here's a breakdown of the difference between procurement and purchasing:

Procurement:

Procurement is a broader concept that encompasses the entire process of acquiring goods, services or works that an organization needs to fulfill its objectives. It involves strategic planning, sourcing, contracting, supplier relationship management, and the overall management of the supply chain. Procurement focuses on obtaining the right resources, from the right sources, at the right time, and at the right cost to support organizational goals and operations.

Key aspects of procurement include:

1. Strategic sourcing: Identifying potential suppliers, evaluating their capabilities, negotiating contracts, and establishing long-term partnerships.
2. Supplier management: Evaluating supplier performance, fostering collaboration, and managing relationships to ensure quality, reliability, and delivery of goods and services.
3. Risk management: Assessing and mitigating potential risks associated with suppliers, market conditions, and supply chain disruptions.
4. Value optimization: Identifying cost-saving opportunities, improving operational efficiency, and maximizing the overall value derived from procurement activities.

Purchasing:

Purchasing, on the other hand, is a subset of the procurement process. It specifically refers to the transactional aspect of acquiring goods or services from chosen suppliers. Purchasing focuses on executing the actual buying process, including requisition processing, requesting quotes, issuing purchase orders, tracking deliveries, and making payments.

Key aspects of purchasing include:

1. Requisition processing: Receiving and reviewing purchase requisitions from various departments or stakeholders within an organization.

2. Supplier selection: Evaluating supplier bids or quotations based on factors such as price, quality, delivery times, and terms and conditions.
3. Purchase order issuance: Creating and sending purchase orders detailing the requested items, quantities, prices, and delivery dates to selected suppliers.
4. Receipt and payment: Receiving the purchased goods or services, verifying their quality and quantity, and processing payments to suppliers.

In summary, procurement encompasses the entire lifecycle of acquiring goods and services, including strategic planning, sourcing, supplier management, and supply chain optimization. Purchasing, on the other hand, focuses solely on the transactional aspects of buying goods or services from selected suppliers.

Procurement Process

The procurement process in a typical EPC project execution in the oil and gas sector involves several key steps. Here is a step-by-step methodology:

1. Identify Procurement Requirements: The first step is to identify the procurement requirements specific to the project. This includes determining the materials, equipment, and services needed and creating a detailed procurement plan.
2. Prequalification of Suppliers: The next step is to prequalify potential suppliers. This involves assessing their capabilities, experience, financial stability, and compliance with project requirements and regulations. The prequalification process helps in shortlisting suppliers who can meet the project's needs.
3. Tendering and Bidding: Once suppliers are prequalified, the project owner or procurement team initiates the tendering process. This involves issuing a Request for Proposal (RFP) or Request for Quotation (RFQ) to the shortlisted suppliers. The

suppliers then submit their bids, including pricing, technical specifications, delivery terms, and contract conditions.

4. Bid Evaluation: After receiving the bids, the project team evaluates them based on various criteria. This includes assessing the technical compliance, commercial terms, pricing, quality, and delivery capabilities of the suppliers. The evaluation may also consider factors such as supplier track record, financial stability, and reputation in the industry.

5. Supplier Selection: Based on the bid evaluation, the project team selects the most suitable supplier(s) for each procurement requirement. The selection criteria generally include a combination of technical compliance, quality, pricing, delivery capabilities, and supplier's ability to meet project specifications.

6. Negotiation and Contracting: Once the suppliers are selected, the project team initiates negotiations with the preferred supplier(s) to finalize the terms and conditions of the contract. This includes negotiating pricing, delivery timelines, payment terms, warranties, and other contractual obligations. Once the negotiations are completed, a formal contract is signed with the selected supplier(s).

7. Purchase Order and Expediting: After the contract is signed, a purchase order is issued to the supplier(s) detailing the specific requirements, quantities, delivery schedule, and other relevant terms and conditions. The project team then monitors and expedites the procurement process to ensure timely delivery of materials and equipment as per the project schedule.

8. Inspection and Quality Control: As the materials and equipment are received at the project site, they undergo inspection and quality control to ensure they meet the specified standards and project requirements. Non-conforming items are identified and dealt with according to the contract terms and quality control procedures.

9. Supplier Relationship Management: Throughout the procurement process, maintaining a strong relationship with suppliers is essential. Regular communication, collaboration, and performance evaluations help ensure the suppliers meet their obligations, resolve any issues, and contribute to the successful execution of the EPC project.

By following this step-by-step methodology, the procurement process in an EPC project in the oil and gas sector can be effectively managed, enabling the timely and cost-effective acquisition of materials, equipment, and services required for the project.

Expediting

Expediting in EPC project execution in the oil and gas sector refers to the process of closely monitoring and proactively managing the progress of procurement activities to ensure the timely delivery of materials and equipment. It involves coordinating with suppliers, tracking orders, and taking necessary actions to expedite or accelerate the procurement process.

The expediting process typically includes the following activities:

1. Monitoring Order Status: Expediting requires regular monitoring of the status of purchase orders placed with suppliers. This involves tracking orders at various stages of the procurement cycle, such as order acknowledgment, production schedule, quality inspection, packaging, and shipment.

2. Communication with Suppliers: Expediting involves maintaining close communication with suppliers to obtain timely updates on the progress of orders, potential delays, or any changes in delivery schedules. Regular follow-ups and open lines of communication help ensure suppliers prioritize the project's requirements.

3. Identifying Bottlenecks: Expeditors are responsible for identifying bottlenecks or potential issues that could cause delays in the

procurement process. This may include identifying production or manufacturing delays, material shortages, capacity constraints, or any other factors that could impact the timely delivery of materials and equipment.

4. Problem Resolution: In the expediting process, if any issues or delays are identified, expeditors work closely with suppliers and other relevant stakeholders to address and resolve them. This may involve negotiating revised delivery schedules, sourcing alternatives from different suppliers, or finding solutions to mitigate potential risks.

5. Prioritizing Critical Items: Expediting requires prioritizing critical items that are crucial for project progress. These may include long-lead items, specialized equipment, or materials required for critical path activities. By giving special attention and dedicating resources to these items, expeditors help ensure they are delivered on time to avoid project delays.

6. Escalation and Reporting: Expeditors play a crucial role in escalating critical issues to project management or senior stakeholders when required. They provide regular status reports, highlighting potential risks or delays, and propose appropriate mitigation measures to keep the project team informed.

The primary objective of expediting is to minimize delays, ensure on-time delivery of materials and equipment, and align the procurement process with the project schedule and overall objectives. Having effective expediting procedures in place helps to mitigate risks, optimize project timelines, and reduce the potential impact of delays or disruptions on EPC projects in the oil and gas sector.

Logistics

Logistics is a critical component of Supply Chain Management (SCM) in EPC project execution in the oil and gas sector. It involves the planning, coordination, and management of the flow of materials, equipment, and services from suppliers to the project site. Effective logistics strategies ensure the availability of resources at the right time, in the right quantity, and at the right location to support the project schedule and objectives.

The following are key aspects of logistics in SCM for EPC projects in the oil and gas sector:

1. Transportation and Routing: Logistics involves selecting the most appropriate transportation mode (e.g., road, rail, air, sea) for the materials and equipment being transported. It also includes determining the optimal routing, taking into consideration factors such as distance, cost, accessibility, and any regulatory or operational restrictions.

2. Customs and Trade Compliance: In the oil and gas sector, EPC projects often involve international suppliers, leading to the need for compliance with customs and trade regulations. Logistics professionals need to ensure proper documentation, customs clearance, and compliance with import/export requirements to prevent delays and unnecessary costs.

3. Warehousing and Inventory Management: Logistics includes the management of warehouses or storage facilities to store materials and equipment before they are needed at the project site. Efficient inventory management practices, such as demand forecasting, stock monitoring, and material handling, are crucial to minimize excess stock, avoid stockouts, and optimize working capital.

4. Material Handling and Packaging: Logistics professionals must coordinate the proper handling and packaging of materials and equipment to ensure their safety during transportation, storage, and installation. This may involve using specialized packaging methods, like heavy-duty crates or protective coatings, to safeguard delicate or valuable items.

5. Track and Trace: Tracking the movement of materials and equipment throughout the supply chain is vital. Advanced technologies, such as barcoding, Radio Frequency Identification

(RFID), or Global Positioning System (GPS) tracking, enable real-time visibility of shipments, helping to manage and monitor their status, anticipate any potential disruptions, and ensure on-time delivery.

6. Reverse Logistics: In EPC projects, there is often a need to manage the return and disposal of materials and equipment at the project's end. Logistics professionals must handle reverse logistics processes, including arranging for the return of unused or defective items, managing disposal in compliance with environmental regulations, and handling any warranty claims or refunds.

7. Risk Management: Logistics encompasses addressing and mitigating risks associated with transportation and supply chain disruptions. This includes developing contingency plans, identifying alternate logistics routes or suppliers, and implementing measures to minimize the impact of unforeseen events such as natural disasters, port closures, labor strikes, or geopolitical conflicts.

Effective logistics management in EPC projects in the oil and gas sector ensures that materials and equipment are delivered on time, in the right condition, and the right quantity. It helps minimize project delays, cost overruns, and disruptions, thus enabling the successful execution of EPC projects in the oil and gas sector.

Vendor (Supplier) management

Vendor management in the EPC (Engineering, Procurement, and Construction) sector of the oil and gas industry involves the coordination and oversight of relationships with vendors, suppliers, and subcontractors throughout the project lifecycle. It encompasses various activities aimed at ensuring the timely delivery of high-quality materials, equipment, and services required for successful project execution.

1. Vendor Prequalification: Vendor management begins with the prequalification process, where potential vendors are evaluated based on criteria such as technical capabilities, financial stability, quality assurance practices, safety records, and compliance with applicable regulations and industry standards. The objective is to identify and select reliable and qualified vendors.

2. Vendor Selection and Contracting: Once vendors are prequalified, the vendor selection process takes place. This involves analyzing bids, proposals, and technical capabilities to determine the most suitable vendor for each scope of work. Contract negotiations are then conducted to define terms, conditions, deliverables, pricing, and other contractual obligations.

    *   Request for Quotation
    *   Technical bid Evaluation
    *   Commercial bid evaluation

3. Vendor Performance Monitoring: Throughout the project, vendor performance is closely monitored to ensure adherence to contract specifications, schedules, and quality requirements. Key performance indicators (KPIs) may be established to track and evaluate vendor performance in areas such as on-time delivery, quality control, safety records, and responsiveness to issues.

4. Relationship Management: Effective vendor management incorporates relationship management practices. This involves maintaining open and regular communication with vendors, fostering collaborative relationships, and resolving any disputes or issues that may arise promptly and professionally.

5. Quality Assurance and Documentation: Vendor management includes establishing quality assurance processes for incoming materials, equipment, and services. This may involve conducting inspections, quality checks, and audits to ensure compliance with project specifications and relevant industry standards. Proper documentation and record-keeping of vendor-related

activities are maintained for future reference and project audit purposes.

6. Contractual Compliance and Payment Management: Vendor management involves overseeing compliance with contractual terms and conditions, including payment schedules, invoicing, and milestones. Regular review and verification of invoices against achieved deliverables and project progress are part of this process.

7. Lessons Learned and Continuous Improvement: Vendor management also includes capturing and applying lessons learned from vendor-related activities. Improvement opportunities identified during vendor management are used to enhance future vendor selection and management processes.

Efficient vendor management in the EPC oil and gas sector ensures the availability of quality materials, equipment, and services at the right time and place. It contributes to project success, cost control, risk mitigation, and the overall delivery of projects following client requirements and industry standards.

Inspection of procured materials and equipment

Depending on the Criticality rating of the equipment the inspection levels are determined.

Commonly used level of inspection as follows:

Level 0: Documentation requirements only; no vendor inspection required.

Level 1: Only final inspection including release for shipment is required before shipping.

Level 2: Includes pre-inspection meetings, one or more unspecified "in progress" surveillance visit/s, all witness and hold points, final inspection, and release for shipment.

Level 3: Includes pre-inspection meetings, one or more unspecified "in progress" surveillance visits, all witness and hold points, final inspection, and release for shipment. Inspections shall be done regularly (daily, weekly, or bi-weekly).

Level 4: Resident inspector continually monitors the work.

Inspection and Test plan (ITP)

ITP shall provide:

Detailed inspection activities and the required tests which include, as a minimum, inspection hold, witness, review points, and inspection frequency.

Sampling plan criteria for inspectable bulk material which shall be determined based on company standards and international code.

Clearly defined roles and responsibilities for each inspection activity.

Procurement Risk Mitigation Guidelines

Risk Level	Contractor Inspector	Company Monitoring	Senior Inspector	Purchase Order Focused Assessment											
				Inspection			PMT			Proponent			Technical SME		
				30%	60%	90%	30%	60%	90%	30%	60%	90%	30%	60%	90%
Critical	☐	☐	☐	☐	☐	☐	☐	☐	☐		☐			☐	☐
High	☐	☐	☐		☐	☐		☐	☐						☐
Medium	☐	☐				☐									
Low	☐														

Remote inspection

Post-COVID (2020) pandemic remote inspection has become common in the EPC industry. This saves time and money and can also be recorded as a video for future reference.

> ### TITBITS
>
> Often we get mixed up using these terms. APPENDIX, ANNEXURE, and ATTACHMENT
>
> Annexure, appendix, and attachments are terms often used in documents, reports, and publications to refer to additional material or supplementary information. While they are similar in that they all serve to provide extra information, they have slightly different connotations and usage:

Annexure:

Annexure is a term commonly used in legal, official, or formal documents, particularly in India and some other Commonwealth countries.

An annexure typically contains supporting documents, additional details, or supplementary information that is directly related to the main document. It is usually numbered or labeled as Annexure A, Annexure B, etc., and referenced within the main document.

Annexures are considered an integral part of the main document and are often used in legal contracts, agreements, and government reports.

Appendix:

An appendix is a section that appears at the end of a document, report, book, or research paper. It is used to provide supplementary or additional information that enhances the understanding of the main content but is not essential for the main narrative.

Appendices are typically numbered or labeled with letters (e.g., Appendix A, Appendix B) and contain charts, graphs, tables, detailed data, or other materials that support or complement the main text.

Appendices are commonly used in academic and research papers, technical manuals, and books to maintain the flow of the main text while offering readers the option to delve into additional details if needed.

Attachments:

Attachments are additional documents or files that are physically or electronically attached to the main document, email, or message.

In the context of emails and electronic communications, attachments are files (e.g., documents, images, spreadsheets) that are sent along with the main message. They are not typically incorporated into the main content but are separate files linked to the message.

In a business or legal context, attachments might refer to separate documents or forms that are submitted or included with a main document, such as an application form with supporting documents attached.

In summary, while an annexure and appendix are sections within a document that provide additional information, an annexure is more common in legal and formal documents, while an appendix is a general term used in various types of reports and publications. Attachments, on the other hand, are separate files or documents that accompany a main document or message, often used in email communication or submissions.

I am not intending to attach anything to this handbook... that's a good decision. isn't it?

Packaged equipment

CHAPTER 7

Construction

Construction management

Construction Management in the EPC (Engineering, Procurement, and Construction) sector of the oil and gas industry is a critical function that ensures the successful execution of construction activities for various EPC projects. It involves coordinating, planning, and overseeing all aspects of the construction phase, from mobilization to completion. Here is a brief overview of Construction Management in the oil and gas sector:

1. Planning and Scheduling: Construction Management begins with the development of a detailed construction plan and schedule.

This includes identifying project activities, defining their sequence, estimating resource requirements, and establishing milestones. Planning and scheduling help optimize the utilization of resources, manage dependencies, and ensure timely project completion.

2. Procurement and Supply Chain Management: Construction Management involves coordinating with the procurement team to ensure timely procurement of materials, equipment, and services required for construction. Effective supply chain management ensures that the necessary resources are available when needed, minimizing project delays and maximizing construction efficiency.

3. Contractor Management: Construction Management oversees the selection, engagement, and management of contractors and subcontractors involved in construction activities. This includes negotiating agreements, monitoring contractor performance, ensuring compliance with safety and quality standards, and resolving any issues that may arise during construction.

4. Health, Safety, and Environmental (HSE) Management: Construction Management prioritizes the implementation and enforcement of robust health, safety, and environmental practices. It includes developing and implementing safety plans, conducting regular inspections, and ensuring regulatory compliance to protect the well-being of workers and minimize environmental impact.

5. Quality Assurance and Quality Control: Construction Management ensures the implementation of quality assurance and quality control processes to maintain the highest construction standards. This involves conducting inspections, quality audits, and testing at various stages to verify compliance with project specifications, codes, and industry standards.

6. Construction Progress Monitoring: Construction Management closely monitors construction progress, comparing actual progress

against the planned schedule, milestones, and deliverables. This includes conducting regular site visits, reviewing progress reports, tracking key performance indicators, and identifying and addressing any deviations from the plan.

7. Coordination and Communication: Construction Management facilitates effective communication and coordination among different project stakeholders. This involves coordinating with engineering teams, clients, suppliers, and other stakeholders to ensure smooth project execution, resolve conflicts, and address project-related issues promptly.

8. Change Management: Construction Management handles and manages changes that may arise during construction, such as design modifications, scope adjustments, or unforeseen circumstances. It involves assessing the impact of changes, obtaining necessary approvals, adjusting the construction plan, and managing associated risks and costs.

Effective Construction Management in the oil and gas sector ensures the successful completion of projects on time, within budget, and in compliance with quality and safety standards. It requires strong leadership, excellent communication skills, and a thorough understanding of construction processes and industry best practices. By efficiently managing construction activities, Construction Management contributes to the overall success of EPC projects in the oil and gas industry.

Quality Control during fabrication/construction at the site

During the fabrication and installation phases of EPC (Engineering, Procurement, and Construction) projects in the oil and gas sector, various quality control activities are performed to ensure the compliance of materials, equipment, and construction processes with specified standards and requirements. Here is a brief detailing of the quality control activities typically conducted during fabrication and installation in the oil and gas sector:

1. Incoming Material and Component Inspection: Quality control starts with the inspection of incoming materials, components, and equipment. This involves verifying that delivered items conform to the specified requirements, conducting visual inspections, checking documentation and certifications, and performing necessary tests or sampling to ensure quality and compliance.

2. Welding and Fabrication Inspection: Quality control activities involve monitoring and inspecting welding and fabrication processes. This includes ensuring proper welding procedures are followed, conducting visual and non-destructive testing (NDT) examinations to detect any defects or discontinuities, and verifying adherence to welding specifications and standards.

3. Dimensional and Fit-Up Inspection: Quality control personnel perform dimensional and fit-up inspections to confirm that fabricated components match the required dimensions, and tolerances, and fit properly within the overall structure. This includes measurements, assessments of alignment, and verifying compliance with design drawings and specifications.

4. Non-Destructive Testing (NDT): NDT techniques, such as ultrasonic testing, radiography, magnetic particle inspection, liquid penetrant testing, and visual inspection, are commonly employed during fabrication and installation. These tests evaluate the integrity, performance, and quality of materials, welds, and structures without causing damage.

5. Coating and Surface Preparation Inspection: Quality control activities encompass inspecting surface preparation and coating application processes. This involves verifying the appropriate surface preparation methods, ensuring adherence to coating specifications, conducting adhesion tests, and visual inspections, and monitoring environmental conditions to prevent coating defects and achieve desired performance.

6. Documentation and Record-Keeping: Throughout fabrication and installation, quality control personnel maintain detailed documentation and records of inspections, test results, certifications, and compliance records. This documentation helps track quality milestones, ensures traceability, and facilitates auditing and accountability.

7. Final Inspections and Acceptance Criteria: Before components or structures are released for installation or commissioning, final inspections are conducted to verify that all quality requirements have been met. Acceptance criteria are checked against engineering standards, project specifications, and applicable codes to ensure compliance and readiness for the next project phase.

These quality control activities during fabrication and installation in the oil and gas sector aim to maintain high standards of quality, enhance safety, prevent potential risks or defects, and ensure compliance with industry standards and client expectations. Rigorous quality control processes contribute to the overall success and performance of EPC projects in the oil and gas industry.

OPPORTUNITIES FOR ENGINEERS TO UPGRADE IN QUALITY CONTROL

In the oil and gas industry within the EPC (Engineering, Procurement, and Construction) sector, there are several certifications available for inspectors that validate their knowledge, skills, and expertise in conducting inspections and ensuring quality within various areas. Some of the different certifications available for inspectors in the oil and gas industry EPC include:

1. Certified Welding Inspector (CWI): The CWI certification, offered by the American Welding Society (AWS), validates the knowledge and skills of inspectors responsible for inspecting and evaluating welded materials and structures. It is highly relevant for inspectors involved in welding processes within the oil and

gas industry. (CSWIP, UK too provides similar certifications for welding inspectors)

2. Certified Quality Inspector (CQI): The CQI certification, provided by the American Society for Quality (ASQ), demonstrates an individual's competence in quality inspection methods, techniques, and practices. It applies to inspectors involved in quality control and assurance activities within the oil and gas EPC projects.

3. NDT Certifications: Non-Destructive Testing (NDT) certifications verify the proficiency of inspectors in using different NDT techniques to detect and evaluate defects or irregularities in materials and structures. Various NDT certifications are available, such as ASNT Level II or III certifications in methods like ultrasound, radiography, magnetic particle testing, liquid penetrant testing, and visual inspection.

4. API Certifications: The American Petroleum Institute (API) offers certifications relevant to inspectors involved in the oil and gas industry. For example, the API 510 certification focuses on pressure equipment inspection, API 570 focuses on piping inspection, and API 653 focuses on aboveground storage tank inspection.

5. Coating Inspector Certifications: Coating inspector certifications, such as NACE Coating Inspector certifications (e.g., NACE CIP Level 1, 2, or 3) or SSPC Coating Inspector certifications (e.g., SSPC PCI level 1, 2, or 3), validate the knowledge and skills of inspectors responsible for assessing and ensuring proper coating application and corrosion protection measures for structures and equipment in the oil and gas industry. (BGas UK also provides such certifications for painting inspectors.)

6. API Q1 and Q2 Certifications: These certifications focus on quality management systems within the oil and gas industry.

API Q1 applies to organizations involved in design and manufacturing, while API Q2 is tailored for organizations involved in service supply chains. Inspectors with these certifications demonstrate their knowledge of quality assurance practices and the ability to conduct inspections in line with API standards.

Painting and Coating

As per the guidelines set by BGas (The British Gas Association), the painting process for industrial vessels follows a systematic procedure, ensuring high-quality coatings and adherence to industry standards. Here is a simplified overview of the painting process as per the BGas guidelines:

1. Surface Preparation:

   - Thoroughly clean the vessel's surface, removing contaminants, grease, oil, rust, and any existing coatings.
   - Perform surface profile measurements to ensure the desired surface roughness is achieved.
   - Repair any surface defects, such as pits or cracks, using appropriate methods.

2. Coating Selection:

   - Choose coatings that are compatible with the vessel's substrate, location, and environmental conditions.
   - Consider factors like corrosion resistance, UV protection, chemical resistance, and temperature stability in coating selection.

3. Mixing and Application:

   - Carefully follow manufacturer specifications for mixing ratios and pot life of the chosen coating materials.

- Use suitable application methods such as airless spray, brush, and roller, or mechanized systems for consistent and uniform coverage.
- Apply multiple coating layers, ensuring proper drying or curing time between each coat as specified by the manufacturer.

4. Quality Control:

- Regularly inspect and measure the Wet film thickness (WFT) and dry film thickness (DFT) using appropriate instruments to ensure compliance with specifications.
- Conduct adhesion tests to assess the bond strength between the coating and the substrate.
- Check coating appearance, including color consistency, gloss, and surface finish.

5. Environmental Considerations:

- Ensure proper ventilation and airflow during coating application to minimize the inhalation of harmful vapors and dust.
- Take necessary precautions to prevent contamination or pollution of surrounding areas while handling and disposing of coating materials.

6. Documentation and Record Keeping:

- Maintain detailed records of coating materials used, batch numbers, application procedures, and inspection results.
- Document any deviations from the specified procedure, corrective actions, and change orders.

7. Maintenance and Inspection:

- Regularly monitor the coated surface for signs of degradation, such as corrosion, peeling, or blistering.
- Schedule routine inspections and touch-up maintenance to address any coating defects, ensuring long-term coating performance.

It is important to note that the specific painting process may vary depending on the vessel's material, project requirements, and applicable industry standards. Following the guidelines set by BGas or any relevant regulatory body helps ensure the achievement of high-quality coatings that provide effective protection and durability for industrial vessels.

Inspection and testing of painted surface

As per the guidelines outlined by BGas (The British Gas Association), the inspection and testing of paint coatings involve several key procedures to ensure the quality and integrity of the applied coatings. Here is an overview of the inspection and testing process as per BGas:

1. Visual Inspection:

   - Conduct a visual inspection of the coated surface to evaluate the general appearance, uniformity, and any visible defects or irregularities.
   - Ensure the coating thickness is visually uniform and consistent with the specified requirements.
   - Check for the presence of coating holidays (areas where the coating is incomplete or missing).

2. Dry Film Thickness (DFT) Measurement:

   - Measure the dry film thickness using non-destructive testing methods such as magnetic gauges, eddy current gauges, or ultrasonic thickness gauges.
   - Verify that the measured coating thickness complies with the specified requirements and tolerances.

3. Adhesion Testing:

   - Perform adhesion testing to assess the bond strength between the coating and the substrate.
   - Use appropriate methods such as cross-cut adhesion testing or pull-off testing to evaluate adhesion.

- Assess the degree of adhesion based on established industry standards and acceptance criteria.

4. Wet Film Thickness (WFT) Measurement:

- Measure the wet film thickness during the application of the coating using wet film thickness gauges.
- Ensure that the applied wet film thickness adheres to the specified range and does not exceed recommended limits.

5. Surface Cleanliness and Contamination Testing:

- Inspect the surface for cleanliness and freedom from contaminants that may affect coating adhesion and performance.
- Perform testing methods like water break test, solvent rub test, or swab testing to determine the presence of contaminants.

6. Holiday Detection Testing:

- Conduct holiday detection testing to identify any discontinuities or gaps (holidays) in the coating system that may allow corrosion to occur.
- Utilize methods such as low voltage wet sponge or high voltage spark testing to detect and locate coating holidays.

7. Documenting and Recording:

- Maintain clear and comprehensive records of inspection and testing activities, including test results, methods used, and any deviations from specified requirements.
- Document any corrective actions taken to address identified issues or non-compliances.

Inspection and testing of paint coatings help ensure that the applied coatings meet the intended specifications and provide adequate protection to the surface. Following the guidelines provided by BGas or other relevant industry standards enables a reliable assessment of the coating quality and adherence to corrosion prevention measures.

Paint Defects

As per the guidelines provided by BGas (The British Gas Association), several major paint defects can occur during coating application. These defects are classified into various categories based on their appearance and underlying causes. Here are some of the major paint defects outlined by BGas:

1. Blistering:

   - Blistering appears as raised bubbles or blisters on the surface of the coating.
   - This defect is often caused by trapped moisture, air, or other gases during coating application or improper surface preparation.
   - Inadequate ventilation or high humidity can contribute to blister formation.

2. Cracking:

   - Cracking refers to the formation of visible cracks or fissures on the coating surface.
   - It can occur due to excessive film thickness, inadequate flexibility of the coating, or structural movement.
   - Insufficient drying or curing time between coats can also lead to cracking.

3. Sagging and Running:

   - Sagging and running occur when the coating forms irregular drips, streaks, or runs during the application.
   - These defects may result from excessive coating thickness, gravity effects, improper technique, or insufficient drying time between coats.

4. Orange Peel:

   - The orange peel effect is characterized by a textured surface resembling the skin of an orange.

- It can be caused by improper spray technique, incorrect or inconsistent viscosity of the coating material, or improper air pressure during application.

5. Poor Adhesion:

- Poor adhesion refers to a lack of bonding between the coating and the substrate.
- It can be caused by inadequate surface preparation, the presence of contaminants, incompatible coating systems, or poor-quality control during application.

6. Discoloration:

- Discoloration manifests as uneven color or variations in shade across the coated surface.
- It can be caused by improper mixing or application of the coating, incompatibility with underlying coatings, or exposure to excessive heat or UV radiation.

7. Flaking and Peeling:

- Flaking and peeling occur when the coating starts to detach from the substrate, resulting in visible flakes or patches.
- These defects may arise due to poor surface preparation, inadequate adhesion, insufficient compatibility, or exposure to environmental factors like moisture or temperature fluctuations.

These are just some of the major paint defects that can impact coating quality and aesthetics. Proper surface preparation, careful application techniques, and adherence to manufacturer specifications can help minimize the occurrence of these defects. Regular inspection and quality control measures are essential to detect and address any paint defects during the coating process.

Difference between painting and coating

While painting and coating are often used interchangeably, there are distinct differences between the two terms:

Painting:

Painting refers to the application of a liquid substance, often referred to as paint, onto a surface for aesthetic purposes. It involves using brushes, rollers, or spray equipment to apply pigmented coatings that enhance the appearance of the surface. Painting is commonly associated with residential, commercial, and artistic applications, such as painting walls, furniture, or artwork. The primary goal of painting is to provide visual appeal, change the color or texture of a surface, and provide a protective layer against minor wear and tear.

Coating:

Coating, on the other hand, is a more generalized term encompassing a broader range of applications and purposes. It involves the application of a protective, functional, or decorative layer on a substrate. Coatings can be applied to various surfaces, including metal, concrete, wood, plastics, or fabrics. Unlike painting, coatings are not solely focused on aesthetics but serve specific purposes such as corrosion protection, chemical resistance, UV resistance, insulation, or lubrication. Coating materials can include paints, varnishes, lacquers, enamels, epoxy, powder coatings, or thermal spray coatings.

In summary, while painting is a subset of coating that primarily focuses on aesthetic enhancements, coatings encompass a wider range of functional and protective applications. Painting is typically associated with surface decoration and visual appeal, while coatings serve a broader spectrum of purposes, including corrosion prevention, durability enhancement, heat resistance, or specialized functional requirements.

Three-layer polypropylene coating on a line pipe.

3L PP Coating of a Line Pipe: Enhancing Performance and Protection

In the realm of line pipe coating, 3L PP (Three Layer Polypropylene) coating has gained significant recognition for its exceptional performance in protecting pipelines from corrosion and other environmental challenges. The application of 3L PP coating involves a meticulous process designed to maximize durability, adhesion, and long-term reliability. Let's explore the key aspects of 3L PP coating for line pipes:

1. Pre-Surface Preparation: Before applying the 3L PP coating, the line pipe undergoes thorough surface preparation. This involves cleaning the pipe surface to remove any contaminants, rust, or mill scale. Surface preparation techniques such as abrasive blasting or chemical cleaning are employed to create an optimal surface condition for coating adhesion.

2. Primer Application: To promote adhesion, a primer is applied as the first layer onto the cleaned line pipe surface. The primer enhances bonding between the pipe and the subsequent 3L PP coating layers.

3. Extrusion of Three-Ply Layers: After the primer, the process moves onto the extrusion of the three-ply layers that form the 3L PP coating system. This system consists of a fusion-bonded epoxy (FBE) layer, an adhesive layer, and finally, the outermost polypropylene (PP) layer.

   - Fusion-Bonded Epoxy (FBE) Layer: The FBE layer acts as the bonding agent between the primer and the adhesive. It provides excellent corrosion protection and enhances adhesion between different layers.

   - Adhesive Layer: The adhesive layer facilitates a strong bond between the FBE layer and the outer PP layer. It promotes interlayer adhesion and contributes to the overall integrity of the coating system.

- Polypropylene (PP) Layer: The outermost layer of 3L PP coating is made of polypropylene. This layer provides exceptional resistance against environmental elements, including moisture, chemicals, UV radiation, and mechanical damage. It serves as a protective barrier, maintaining the integrity of the system.

4. Quality Control and Inspection: Throughout the coating process, strict quality control measures are implemented to ensure the coating's integrity and adherence to industry standards. Visual inspections, adhesion tests, and coating thickness measurements are performed to verify the quality and reliability of the 3L PP coating.

5. Performance Testing: Coated line pipes may undergo rigorous performance testing, such as corrosion resistance, impact resistance, and aging tests, to evaluate the coating's durability and effectiveness in real-world conditions.

The 3L PP coating of line pipes offers numerous benefits, including excellent corrosion protection, long-lasting performance, and resistance to a wide range of environmental factors. It provides a cost-effective solution for safeguarding pipelines in diverse industries such as oil and gas, water, and chemical transportation.

With its advanced material properties and careful application, 3L PP coating enhances the longevity and reliability of line pipes, ensuring the safe and efficient transport of fluids. It exemplifies the ongoing pursuit of innovative coating solutions that prioritize protection, performance, and sustainability in the dynamic world of pipeline infrastructure.

GRE pipeline

Constructing a GRP (Glass Reinforced Plastic) pipeline involves several key steps to ensure proper installation, durability, and performance. Here is a simplified overview of the process for constructing a GRP pipeline:

1. Site Preparation: Before pipeline construction begins, the site is prepared by clearing and leveling the ground, ensuring a suitable foundation for the pipeline installation.

2. Trench Excavation: A trench is excavated along the planned pipeline route. The trench dimensions and depth should adhere to project specifications and regulatory requirements, considering factors such as soil conditions and pipeline depth.

3. Bedding Preparation: The trench is prepared with suitable bedding material to provide support and stability for the GRP pipeline. This may involve adding sand or other suitable materials to create a level and consistent surface.

4. Pipeline Assembly: The GRP pipeline sections, typically in the form of prefabricated pipes or segments, are brought to the site. These sections are joined together using appropriate methods such as filament winding, butt fusion, or adhesive bonding, depending on the specific type of GRP pipeline being installed.

5. Pipeline Installation: The assembled GRP pipeline sections are carefully lowered into the prepared trench. Special attention is given to maintaining alignment and preventing damage during the installation process. Pipe supports and alignment guides may be used as necessary to ensure proper positioning.

6. Joint and Connection Installation: Joints and connections between pipeline sections are installed using methods specific to GRP pipes, such as mechanical couplings, adhesive bonding, or flanged connections, depending on project requirements and standards.

7. Backfilling: Once the pipeline is successfully installed, the trench is backfilled with suitable material in layers, carefully compacting and ensuring proper compaction to provide additional support and stability to the pipeline.

8. Testing and Inspection: The constructed GRP pipeline undergoes thorough testing and inspection to ensure its integrity, tightness, and compliance with industry standards and project

specifications. Common tests include hydrostatic pressure testing to verify the pipeline's strength and leak resistance.

9. Restoration: The construction site is restored and brought back to its original condition, including regarding the land and reinstating any disturbed surfaces, if necessary.

It is important to note that construction procedures may vary depending on project-specific requirements, the type of GRP pipeline being installed, and applicable regulations. Following proper installation procedures, quality control measures, and adherence to industry standards are essential for the successful construction of a GRP pipeline.

Above-ground and underground pipeline construction

1. Visibility and Land Use:

- Above Ground: AG pipelines are visible and located on the surface, often running alongside roads, railways, or through open landscapes. They may require more land and may be subject to specific regulations regarding their proximity to population centers or sensitive areas.
- Underground: UG pipelines are buried beneath the surface, minimizing visual impact and allowing for land use above the pipeline. They are typically located away from public view and may require less land for construction.

2. Construction Method:

- Above Ground: AG pipelines are constructed using traditional open-cut trenching methods, where a trench is excavated, and the pipeline is laid and installed within it. This method allows for easier access during construction and maintenance.
- Underground: UG pipelines can be installed using trenchless technology, such as horizontal directional drilling (HDD) or micro tunneling, which minimizes disruption to the surface and existing infrastructure. This method is utilized when avoiding

surface disturbances is a priority or when navigating obstacles during construction.

3. Protection and Environmental Considerations:

- Above Ground: AG pipelines require protective coatings and wrapping to safeguard against corrosion and external factors such as weather and physical damage. They may also require additional safety measures to protect the pipeline from accidental impact or vandalism.
- Underground: UG pipelines benefit from natural protection provided by the soil, reducing vulnerability to external elements. However, they still require corrosion protection, cathodic protection systems, as well as monitoring and leak detection systems to ensure the integrity of the buried pipeline.

4. Land Restoration:

- Above Ground: AG pipelines may require land clearing and modification, potentially impacting vegetation and surface features. Restoration efforts focus on reinstating the land to its original condition or as agreed upon with landowners.
- Underground: UG pipelines involve minimal surface disturbance, allowing for easier land restoration and minimal impact on vegetation and surface features. Reclamation efforts largely focus on erosion control and soil stabilization.

5. Monitoring and Maintenance:

- Above Ground: AG pipelines are often subjected to visible inspection and regular above-ground maintenance activities, such as visual checks, painting, or repair work. Monitoring for leaks and corrosion is more accessible and easier to detect.
- Underground: UG pipelines require specialized underground inspection tools, such as smart pigs and remote sensing technologies, to monitor the integrity of the buried pipeline.

These inspections typically occur less frequently and may require excavation in certain instances.

6. Regulatory Considerations:

- Above Ground: AG pipelines may have specific regulations regarding their visibility, safety, and proximity to population centers or environmentally sensitive areas. Additional measures may need to be implemented to ensure public safety and compliance with applicable regulations.
- Underground: UG pipelines are subject to similar safety and regulatory requirements as AG pipelines but may have additional considerations related to construction practices, environmental impact, and land use.

These key differences highlight the contrasting characteristics and considerations between above-ground and underground pipeline construction, emphasizing factors such as visibility, construction methods, protection measures, land restoration, monitoring, and regulatory requirements.

Onshore pipelines vs subsea pipelines

Onshore pipelines and subsea pipelines differ primarily in their location and construction methods. Here are the key differences between onshore pipelines and subsea pipelines:

1. Location:

- Onshore Pipelines: Onshore pipelines are located on land, typically running across vast distances and various terrains, including deserts, mountains, forests, or populated areas.
- Subsea Pipelines: Subsea pipelines are installed beneath the seabed, crossing bodies of water such as oceans, seas, or lakes. They are specifically designed to operate in an underwater environment.

2. Construction Methods:

- Onshore Pipelines: Onshore pipelines are often constructed using either open-cut trenching or trenchless techniques such as HDD (horizontal directional drilling) or micro tunneling. These pipelines are typically laid on or beneath the ground surface and can be accessed relatively easily for construction and maintenance.

- Subsea Pipelines: Subsea pipelines are constructed using specialized vessels and equipment. They involve methods like pipelaying, where sections of the pipeline are welded, coated, and then laid on the seabed using remotely operated vehicles (ROVs) or other installation methods.

3. Environmental Considerations:

- Onshore Pipelines: Onshore pipelines encounter various environmental considerations, including land use, preservation of ecosystems, and the potential impact on local communities. Environmental assessments and regulations play a significant role in determining the pipeline's route and design to minimize ecological impact.

- Subsea Pipelines: Subsea pipelines must consider marine life, seabed ecosystems, and potential impacts on underwater habitats. Environmental studies are conducted to assess and mitigate potential disturbances caused during installation, operation, and maintenance.

4. Maintenance and Access:

- Onshore Pipelines: Onshore pipelines are more accessible for maintenance and repairs since they are located on land. Surface inspections, visual checks, and routine maintenance are relatively straightforward and can be performed without significant challenges.

- Subsea Pipelines: Subsea pipelines require specialized equipment and techniques for maintenance and inspection. ROVs or diving operations are employed to access and inspect the pipeline underwater. Maintenance activities may involve remotely operated tools or the deployment of divers to make necessary repairs or perform interventions.

5. Installation Challenges:

- Onshore Pipelines: Onshore pipelines face challenges related to land acquisition, traversing various terrains or landforms, and dealing with potential obstacles such as existing infrastructure, roads, or rivers.
- Subsea Pipelines: Subsea pipelines encounter challenges specific to underwater construction, including challenges related to pipeline stability, installation in deep water, seabed composition, and working in harsh environmental conditions.

6. Pipeline Protection:

- Onshore Pipelines: Protection measures for onshore pipelines usually involve corrosion prevention coatings, cathodic protection systems, and the implementation of safety measures to safeguard the pipeline from external forces or excavation damage.
- Subsea Pipelines: Subsea pipelines require additional protective measures to counteract the effects of corrosion caused by saltwater and marine environments. They may be coated with specialized materials to ensure resistance to corrosion and damage caused by underwater conditions.

Both onshore and subsea pipelines play critical roles in energy transportation, but these key differences highlight the notable distinctions between their location, construction methods, environmental considerations, accessibility, installation challenges, and protective measures. Understanding these differences is essential for planning, designing, and maintaining pipelines in their respective environments.

Hot tapping

Let's delve deeper into the hot-tapping process and its significance:

Hot tapping is a complex procedure that allows for the creation of connections or branches in a pressurized pipeline or vessel while the system remains operational. This method is widely used in various industries such as oil and gas, water distribution, chemical processing, and many others.

The planning and preparation phase is crucial to ensure a successful hot-tapping operation. Factors such as system pressure, temperature, pipe material, and the desired location for the tap are carefully considered. This information helps determine the necessary equipment, tools, and materials for the specific project.

Once the planning phase is complete, specialized hot tap equipment is selected. This can include a hot tap machine, suitable fittings, welding equipment, and isolating methods such as hot tapping valves or line stop fittings. These tools are essential for safely and effectively creating the connection in the live system.

Before the hot tapping process can begin, thorough preparation is crucial. The work area is cleaned, and any coating or insulation on the pipe surface is removed. The hot tap machine is then aligned and attached securely to the pipeline. Preparation for welding involves beveling the pipe and ensuring proper fit-up of the fitting and welding components.

The welding phase is a critical step in hot tapping. Skilled welders perform the connection using controlled welding techniques to ensure the new branch possesses strength, integrity, and reliability. Welding crews work closely together to create a robust connection that can withstand the operational pressures of the system.

Once the welding is complete, the hot tap machine is activated to drill into the main pipeline, creating a hole through which the new branch line can be installed. This process is carefully controlled to minimize disruption and maintain system integrity.

Following the installation of the branch line, various tests are conducted to ensure the quality and effectiveness of the hot tapping operation. Non-destructive testing methods are employed to assess the joint integrity, pressure tests are carried out to assess the strength of the connection, and leak detection measures are implemented to confirm the absence of any leaks.

Hot tapping offers significant benefits to industries, including cost and time savings, reduced system downtime, and minimal disruption to ongoing operations. This method allows for necessary modifications, repairs, or expansions to be made without completely shutting down the entire system. By adhering to proper procedures, and safety protocols, and employing skilled professionals, hot tapping ensures the integrity, reliability, and functionality of the pipeline or vessel while enabling essential changes and improvements to be implemented efficiently and effectively.

Pipeline pigging

The pipeline pigging process is an essential maintenance procedure used in the oil, gas, and liquid transportation industries. It involves the use of specialized devices called pipeline pigs, which are cylindrical or disc-shaped tools inserted into the pipeline to perform various functions.

1. Preparatory Phase:

Before pigging, the pipeline undergoes a thorough preparatory phase. This typically involves conducting a pipeline inspection to assess its condition and identify any potential issues. Cleaning and preparation activities take place to remove debris, sediment, or product buildup that may have accumulated over time.

2. Pig Insertion:

Once the pipeline is prepared, a specialized device called a pipeline pig is inserted into the line. Pipeline pigs come in various designs, such as solid or foam discs, brushes, or spheres. They are usually propelled through the pipeline using the flow of the product or by applying pressure differentials.

3. Cleaning and Debris Removal:

Cleaning pigs are commonly used to remove accumulated debris, sediment, and other contaminants from the pipeline walls. These pigs are equipped with scraping discs, brushes, or other mechanisms that scrape off debris during their passage. In some cases, chemical agents are injected into the line to facilitate cleaning, dissolving deposits, and improving pipeline flow.

4. Inspection and Testing:

Inspection pigs, also known as intelligent or smart pigs, are employed to assess the condition of the pipeline. Equipped with advanced sensors and instruments, these pigs collect valuable data as they travel through the pipeline. They can detect defects such as corrosion, cracks, or wall thickness abnormalities. The gathered data helps identify potential integrity issues and enables informed decisions regarding maintenance or repairs.

5. Maintenance and Repair:

If abnormalities or defects are detected during the inspection process, maintenance pigs or specialized pigs may be deployed to address specific issues. These pigs can be outfitted with tools for maintenance activities like welding, coating applications, or even deploying internal clamps for repairing localized damage.

6. Pig Recovery:

After the pigging operation is completed, the pipeline pigs are safely recovered. Pig traps or retrieval devices are used to capture and extract the pigs from the pipeline. This ensures that no debris or equipment is left behind, which could potentially cause damage or impede future operations.

The pipeline pigging process is crucial for maintaining overall pipeline integrity, optimizing product flow, and ensuring operational safety. Regular pigging operations help prevent blockages, improve product quality, reduce the risk of corrosion, and enable proactive maintenance planning. By implementing effective pigging programs, operators can maximize the lifespan, efficiency, and reliability of their pipeline systems.

Scaffolding

Scaffolding plays a crucial role in construction, providing a safe working platform for workers to access elevated areas and perform tasks efficiently. Here's a brief overview of scaffoldings used in construction:

Scaffolding Types:

There are various types of scaffoldings used depending on the nature of the construction project. Common types include:

1. Tube and Coupler: This traditional scaffolding system utilizes steel tubes and couplers to create a modular framework. It offers flexibility and versatility, making it suitable for various project requirements.
2. System or Modular Scaffold: These pre-engineered scaffold systems consist of standardized components that can be easily assembled and disassembled. They offer faster installation, increased safety, and improved efficiency compared to traditional scaffolding.

3. Suspended Scaffold: Typically used for tasks such as window washing or façade maintenance, suspended scaffoldings are suspended from the roof or a supporting structure, allowing workers to access high-rise building exteriors or inaccessible areas.
4. Mobile Scaffold: These portable scaffoldings are mounted on wheels or casters, providing flexibility and mobility. They are commonly used for tasks with a smaller scope or for projects that require frequent repositioning of the scaffold.

Safety and Regulations:

Safety is paramount in scaffold design and usage. Scaffoldings must meet relevant safety standards and regulations, such as proper stability, load-bearing capacity, and fall protection measures. Regular inspection, maintenance, and adherence to safety protocols are essential to prevent accidents and ensure worker safety.

Components and Accessories:

Scaffoldings consist of various components, including frames, platforms (also known as decks or working platforms), guardrails, and diagonal braces. Accessories such as access stairs, ladders, and toe boards further enhance safety and functionality.

Erecting and Dismantling Scaffoldings:

Scaffoldings are erected following specific procedures to ensure stability and structural integrity. This includes laying the foundation, assembling the scaffold components, securing connections, and adding necessary accessories. Dismantling follows reverse steps, carefully removing components and ensuring safety during the process.

Scaffoldings provide crucial support to construction projects, facilitating access to elevated work areas and ensuring worker safety. With various types available, scaffoldings are customized to meet project requirements. Adherence to safety regulations, proper assembly, and regular inspections are imperative to ensure the stability, functionality, and safe use of scaffoldings on construction sites.

Modularization

Modularization in EPC (Engineering, Procurement, and Construction) construction within the oil and gas industry refers to the practice of designing, fabricating, and assembling pre-engineered modules or skids off-site, which can be transported to the project site for final installation. This approach offers significant advantages in terms of efficiency, cost reduction, schedule compression, and improved quality control.

Key aspects of modularization in the oil and gas industry include:

1. Design and Engineering: Modularization starts with careful planning and detailed engineering. The project is broken down into manageable modules, considering factors such as equipment layouts, piping systems, electrical and instrumentation requirements, structural integrity, and transportation logistics. Collaboration among various engineering disciplines is essential to optimize module design and functionality.

2. Fabrication and Assembly: Once the module designs are finalized, fabrication takes place in controlled off-site facilities called fabrication yards. Skilled workers construct the modules using standardization, proper sequencing, and quality control measures. This controlled environment allows for efficient fabrication, reduced construction time, and enhanced safety.

3. Integration and Testing: After fabrication, modules go through rigorous testing, including functional testing of equipment, pressure testing of piping systems, electrical systems validation, and instrumentation calibration. This ensures that each module meets the required specifications and can seamlessly integrate with other modules during the final installation phase.

4. Transportation and Site Installation: Upon completion of testing, the modules are transported to the project site via land or sea. Careful logistical planning and coordination are critical for the timely and safe delivery of modules. At the site, modules are positioned, aligned, and interconnected to form the complete facility. The interconnection process may involve welding, bolting, or flange connections.

Advantages of Modularization:

Modularization offers several benefits in EPC construction for the oil and gas sector, including:

- Schedule Compression: By parallelizing site preparation and module fabrication, project schedules can be significantly reduced.
- Cost Efficiency: Efficient fabrication processes, economies of scale, reduced labor costs, and improved productivity contribute to overall cost savings.
- Enhanced Quality Control: Fabrication in controlled environments ensures higher quality standards and adherence to specifications.
- Safety Improvements: Off-site fabrication reduces on-site workforce exposure to potential hazards.
- Flexibility and Adaptability: Modularization allows for greater adaptability to project changes, as modules can be easily modified or replaced if necessary.
- Minimal Site Disruption: Off-site fabrication minimizes disruption at the project site, reducing potential impacts on neighboring areas and the environment.

Modularization has revolutionized the construction approach in the oil and gas industry, offering faster project delivery, cost efficiencies, enhanced quality, and improved safety performance. Its widespread adoption continues to transform the construction landscape, driving innovation and efficiency throughout the entire project life cycle.

Advanced work package

In the EPC (Engineering, Procurement, and Construction) industry, an Advanced Work Package (AWP) refers to a detailed and comprehensive package of information for a specific work scope within a project, particularly in the oil and gas sector. It provides a structured and well-defined plan for executing a specific task or portion of the project.

The key features and components of an Advanced Work Package typically include:

1. Scope of Work: A clear and concise description of the work to be performed, specifying the objectives, deliverables, and performance requirements. This includes delineating the boundaries of the work package and its interfaces with other packages or disciplines.
2. Work Breakdown Structure (WBS): A hierarchical breakdown of the work package into manageable components or tasks. The WBS provides a logical structure for planning, scheduling, resource allocation, and reporting progress.
3. Design Documentation: Detailed engineering drawings, specifications, and technical documents related to the scope of work. These documents provide the necessary guidelines for executing the work accurately and in compliance with industry standards, codes, and regulations.
4. Material Requirements: A comprehensive list of materials, equipment, tools, and consumables required for the execution of the work package. This information helps with procurement, ensuring that all necessary resources are available at the appropriate time.
5. Schedule and Milestones: A well-defined timeline that outlines the planned start and end dates for each task within the work package. Milestones are identified to track progress and facilitate coordination between different work packages and stakeholders.

6. Resource Planning: Identification and allocation of the resources needed to execute the work package effectively. This includes staffing requirements, skill sets, equipment, and other resources necessary for successful completion.

7. Quality Assurance and Control: Procedures and guidelines to ensure adherence to quality standards throughout the execution of the work package. This typically involves inspections, testing protocols, and documentation requirements to verify compliance with project specifications.

8. Health, Safety, and Environment (HSE) Considerations: A comprehensive plan addressing health, safety, and environmental risks associated with the work package. This includes protocols, procedures, and guidelines to mitigate potential hazards and ensure compliance with relevant regulations and best practices.

An Advanced Work Package plays a critical role in the efficient execution of projects within the oil and gas sector. It enables effective planning, resource allocation, and progress monitoring, facilitating improved project control and coordination among various stakeholders. A well-prepared and information-rich work package enhances project efficiency, quality, and safety while reducing potential risks and uncertainties.

Sub-contractor management

Subcontractor management in the EPC (Engineering, Procurement, and Construction) sector of the oil and gas industry involves the effective coordination and oversight of subcontractors who perform specific portions of the project scope. It includes activities aimed at ensuring the smooth integration, performance, and delivery of subcontracted work within the project timeline and quality requirements.

1. Subcontractor Selection and Prequalification: Subcontractor management begins with the selection and prequalification process. Potential subcontractors are evaluated based on factors

such as technical expertise, experience, financial stability, safety records, and compliance with relevant regulations and industry standards. Prequalification helps ensure that subcontractors possess the necessary capabilities and meet project requirements.

2. Contracting and Scope Definition: Once subcontractors have been selected, contracts are negotiated and signed, clearly defining the scope of work, deliverables, timelines, quality requirements, and payment terms. These contracts establish the legal framework and expectations for both parties involved.

3. Communication and Collaboration: Effective subcontractor management relies on open and regular communication channels between the EPC contractor and the subcontractors. This includes project kick-off meetings, progress meetings, and routine coordination to ensure alignment of goals, clarify expectations, resolve issues, and foster collaboration throughout the project duration.

4. Performance Monitoring and Quality Control: Subcontractor performance is continuously monitored to ensure adherence to project timelines, quality standards, and safety regulations. Key performance indicators (KPIs) are established and tracked to evaluate subcontractor performance in areas such as productivity, adherence to schedules, safety records, and overall work quality.

5. Change Management: Subcontractor management involves effectively managing any changes or variations that may arise during project execution. This includes assessing the impact of changes, negotiating contractual adjustments, and ensuring proper documentation and communication with subcontractors to avoid disputes and maintain project progress.

6. Risk Management and Safety Compliance: Subcontractor management includes monitoring subcontractor activities to identify and mitigate potential risks to safety, quality, and project performance. This involves conducting safety audits,

providing necessary training, enforcing safety protocols, and ensuring subcontractor compliance with applicable regulations and industry standards.

7. Payment and Contractual Compliance: Effective subcontractor management includes managing payment processes, verifying invoices against achieved milestones or completed work, and maintaining proper documentation for audit and contractual compliance purposes.

8. Lessons Learned and Relationship Building: Subcontractor management incorporates capturing and applying lessons learned from subcontractor-related activities. It also focuses on building long-term relationships with reliable and trusted subcontractors, promoting repeat collaboration, and fostering a positive reputation within the subcontractor community.

Efficient subcontractor management in the EPC oil and gas sector ensures the successful integration and performance of subcontracted work within the broader project scope. It helps mitigate risks, maintain project schedules, optimize resource utilization, and ultimately contribute to the overall delivery of successful EPC projects.

Selection criteria of a construction sub-contractor

The process of selecting a subcontractor for performing specific tasks in an EPC project, such as a refinery construction project, involves several steps and considerations. Subcontractor management is crucial to ensure the smooth execution of project activities and adherence to quality, safety, and schedule requirements.

1. Prequalification and Evaluation: The first step is to identify potential subcontractors through a prequalification process. This involves evaluating their capabilities, experience, financial stability, equipment, and workforce. Prequalification criteria may also consider safety records, past performance, compliance with legal and regulatory requirements, and relevant certifications or licenses.

2. Request for Proposals (RFPs) or Invitation to Bid (ITB): Once a list of prequalified subcontractors is established, an RFP or ITB is issued. These documents include specific project requirements, scope of work, project schedule, and commercial terms. Subcontractors are invited to submit their proposals, detailing their technical and commercial capabilities, proposed methodology, pricing, and any other relevant information.

3. Bid Evaluation and Selection: The project team evaluates the subcontractor proposals, considering technical competence, past performance, proposed approach, cost competitiveness, and alignment with project requirements. A thorough evaluation process ensures that subcontractors are selected based on their suitability for the project.

4. Contract Negotiation and Award: After selecting the preferred subcontractor, negotiations are conducted to finalize the terms and conditions of the subcontract. This includes project scope, deliverables, pricing, payment terms, schedule commitments, safety requirements, quality standards, insurance provisions, and dispute resolution mechanisms. Once both parties agree to the terms, the subcontract is awarded.

5. Subcontractor Mobilization and Onboarding: Once the subcontract is awarded, the subcontractor proceeds with mobilization activities. This involves obtaining necessary permits and approvals, setting up project-specific schedules, establishing project facilities, coordinating logistical requirements, and ensuring compliance with project policies, procedures, and safety protocols.

6. Subcontractor Performance Management: Throughout the project, the subcontractor's performance is closely monitored. This includes regular inspections, progress tracking, and quality assessments to ensure compliance with project requirements. Key performance indicators (KPIs) may be established to measure subcontractor performance in areas such as safety, quality,

schedule adherence, and productivity. Performance reviews and feedback sessions are conducted to address any issues and implement improvement plans as needed.

7. Communication and Collaboration: Effective communication and collaboration are critical in subcontractor management. Regular coordination meetings, progress updates, and site visits are conducted to maintain open lines of communication and align subcontractor activities with project goals. Clear channels of communication ensure prompt resolution of issues, effective change management, and smooth project execution.

8. Contract Administration and Compliance: Contract management activities include reviewing subcontractor invoices, approving progress payments, and ensuring compliance with contractual obligations. Documentation and record-keeping of all subcontract-related correspondence, changes, and variations are maintained. Regular audits may also be conducted to ensure compliance with safety, quality, and regulatory requirements.

9. Dispute Resolution: In case of any conflicts or disputes, well-defined dispute resolution mechanisms, such as mediation or arbitration, are followed to address and resolve differences between the contractor and subcontractor.

Effective subcontractor management ensures that subcontractors contribute positively to project success, maintain high standards of workmanship, adhere to project schedules, comply with safety and quality requirements, and contribute to overall project goals. Regular communication, proactive performance management, and establishing a collaborative working relationship are crucial for successful subcontractor management in an EPC project.

TITBITS

Some of the good books to read for self-development:

- Think and Grow Rich
- The 7 Habits of Highly Effective People
- Man's Search for Meaning
- The Alchemist
- The Power of Positive Thinking
- How to win friends and influence people
- How to stop worrying and start living
- Thirukkural (Original in Tamil)
- Bhagwad Gita

TITBITS

Public speaking

1. **Understand Your Audience**: Tailor your content to the needs and interests of your audience. Understand their background and expectations to make your speech more engaging.

2. **Practice**: Regular practice can help you become more comfortable with your material and delivery. This can include rehearsing in front of a mirror, recording yourself, or practicing in front of friends or family.

3. **Body Language**: Effective use of body language, such as maintaining eye contact, using hand gestures, and standing tall, can help convey confidence and engage your audience. Dress for the occasion, which is important.

4. **Voice Modulation**: Vary your pitch, volume, and speed to keep your audience interested. A monotone voice can

be boring, so try to add some variety to your speech. (Be cautious, do not shout)

5. **Use Visual Aids**: Visual aids, such as slides or props, can help illustrate your points and keep your audience engaged. However, ensure they complement your speech and do not distract from your message. (Never look at the screen and start reading)

6. **Prepare for Questions**: Anticipate questions that may arise from your speech and prepare responses. This can help you feel more confident during the Q&A session. (Even if you are unable to answer, tell that you will comeback with an answer, on that question, soon.)

7. **Manage Nervousness**: Everyone feels nervous before a speech. Techniques such as deep breathing, visualization, or focusing on the message rather than yourself can help manage this nervousness.(Keep a water bottle, handy)

8. **Feedback and Improvement**: Seek feedback after your speech and use it to identify areas for improvement. Continuous learning and improvement are key to becoming a better public speaker.

9. **Read, Read, and Read**: Make reading a routine, and keep taking notes, whenever you come across some good points.

10. **No arguments**: Never ever get into any argument with the audience. Stay calm.

Public speaking is a skill that improves over time. Do not be discouraged by initial difficulties and keep practicing and refining your skills.

CHAPTER 8

Mechanical Completion, Pre-Commissioning and Commissioning

Mechanical Completion

Mechanical Completion is a critical milestone in an EPC (Engineering, Procurement, and Construction) project within the oil and gas sector. It represents the stage when all mechanical and process systems, components, and equipment have been installed, tested, and verified as fully operational.

1. Installation of Equipment and Systems: Mechanical Completion occurs after the successful installation of all mechanical equipment, systems, pipelines, structures, and associated components within the project. This includes process equipment, pumps, compressors, vessels, heat exchangers, instrumentation, control systems, and other mechanical elements required for the project scope.

2. Punch List Clearing: Before Mechanical Completion, a Punch List is created, which consists of identified deficiencies, outstanding works, or pending corrective actions. During the Mechanical Completion phase, the Punch List items are inspected and rectified to ensure that all outstanding issues or deficiencies have been addressed, and the project is ready for commissioning and start-up.

3. Pre-Commissioning Activities: Mechanical Completion is closely linked with pre-commissioning activities. Pre-commissioning refers to a series of tests, inspections, and verifications conducted to ensure that the installed systems and equipment are ready for commissioning. This includes checks on equipment integrity,

cleaning and flushing of piping systems, checking electrical connections, testing instrument performance, and confirming that all safety systems are in place.

4. Documentation and Certification: Mechanical Completion involves thorough documentation and certification of all activities and systems related to the mechanical aspects of the project. This documentation includes test certificates, compliance records, inspection reports, and completion certificates, demonstrating that all mechanical works have been carried out according to specifications, standards, and contractual requirements.

5. Handover Readiness: Mechanical Completion signifies that the project is ready for handover from the EPC contractor to the client or operator. It confirms that the mechanical installation and related activities have been completed as planned, meeting all quality and technical requirements.

Once Mechanical Completion is achieved, the project proceeds to the Commissioning and Start-Up phase, where functional and performance tests are performed, and operations gradually commence.

Mechanical Completion is pivotal in ensuring the readiness and integrity of the mechanical systems and equipment within the oil and gas sector. It signifies the successful completion of mechanical installation, addresses outstanding issues, and serves as a crucial step toward the safe and efficient start-up of the project.

Pre-commissioning

Pre-commissioning in an EPC (Engineering, Procurement, and Construction) project within the oil and gas sector refers to a series of activities carried out before the formal commissioning and start-up of the project's systems, equipment, and facilities. Pre-commissioning activities aim to ensure that all installed components and systems are functioning as intended and are ready for safe and efficient operation.

1. Functional Testing: Pre-commissioning involves the execution of functional tests on individual equipment and systems. These tests verify that each component operates as per design, meets performance criteria, and functions correctly before being integrated into the project as a whole. Examples of functional tests include checking motor rotations, control system functionality, and equipment interlocks.

2. Verification of Instrumentation and Controls: Pre-commissioning activities focus on evaluating the performance and accuracy of instrumentation and control systems. This involves calibration of instruments, verifying sensor readings, testing instrument loops, and ensuring the proper functioning of control valves, safety systems, and alarms.

3. Cleaning and Flushing: Pre-commissioning includes cleaning and flushing processes to remove contaminants, debris, and foreign matter from pipelines, vessels, and equipment. This ensures that the internal surfaces of the system are clean and free from obstructions before commissioning and operation, thus maintaining system integrity and minimizing the risk of operational issues or damage.

4. Piping Integrity and Pressure Testing: Pre-commissioning involves conducting pressure tests to verify the integrity and functionality of piping systems. This includes hydrostatic testing, pneumatic testing, and leak testing to identify and rectify any leaks or defects before the system is commissioned for operation.

5. Dry Runs and Simulation: Pre-commissioning activities may include dry runs and simulations of system operation to assess the reliability, efficiency, and coordination of different components. This process helps identify and rectify any operational issues, validate process sequences, and ensure overall system readiness.

6. Personnel Training: Pre-commissioning provides an opportunity for personnel training and familiarization with the systems

and procedures. This includes training operators on specific equipment, safety protocols, and emergency response procedures, ensuring that they are adequately prepared for the commissioning and start-up phases.

7. Document Preparation: Pre-commissioning involves the preparation and collation of relevant documentation, including test records, inspection reports, compliance certificates, operating manuals, and maintenance procedures. These documents ensure traceability, provide reference materials for operation and maintenance, and demonstrate compliance with regulatory and contractual requirements.

Pre-commissioning plays a crucial role in verifying the functionality, integrity, and safety of the project's systems and equipment before actual commissioning and start-up. It allows for necessary adjustments, rectification of faults, and optimization of systems to ensure a smoother transition to the operational phase. Effective pre-commissioning activities contribute to the successful execution and long-term performance of EPC projects in the oil and gas sector.

Commissioning

Commissioning in an EPC (Engineering, Procurement, and Construction) project within the oil and gas sector refers to the systematic process of testing, verifying, and ensuring that all systems, equipment, and facilities are ready for safe and efficient operation. It involves a series of activities to bring the project to operational status.

1. System Integration: Commissioning involves integrating different systems, components, and units within the project to form a complete operating entity. This includes connecting various mechanical, electrical, and control systems, ensuring their seamless functioning together.

2. Static Testing: The commissioning process begins with static testing to validate the functionality and performance of individual

components and systems. It includes conducting tests such as pressure tests, leak tests, performance tests, and functional checks to ensure that each component operates as per design specifications.

3. Dynamic Testing: Following successful static tests, dynamic testing is conducted to evaluate the overall performance of the integrated systems under operating conditions. This involves simulating real-time scenarios, process flows, and emergencies to verify system response, safety measures, and reliability.

4. Operational Readiness: Commissioning activities ensure that all necessary prerequisites for safe operation are met. This includes verifying the availability of utilities, consumables, and supporting services such as power supply, water, fuel, and waste management. It also involves inspecting and certifying safety systems, emergency shutdown mechanisms, and fire protection measures.

5. Performance Testing: Commissioning includes performance testing to assess the system's compliance with design specifications, performance guarantees, and regulatory requirements. This involves measuring and verifying parameters such as flow rates, pressures, temperatures, capacity, and efficiency to ensure that the system operates as intended.

6. Safety Systems Verification: Commissioning verifies that all safety systems and emergency procedures are in place and functioning effectively. This includes testing safety interlocks, emergency shutdown systems, fire detection and suppression systems, and verifying adherence to safety protocols and regulations.

7. Operator Training: During commissioning, operator training takes place to ensure that the project's operators are fully acquainted with the systems, equipment, procedures, and safety protocols. This includes training on control panels, alarms, operational procedures, maintenance routines, and emergency response measures.

8. Handover to Operations: Once commissioned, the project is considered officially ready for handover to the operations team. This involves transferring project documentation, test records, operating manuals, and maintenance procedures to the operational team for their reference and use.

Commissioning is a crucial phase that ensures that the EPC project is fully functional, reliable, and compliant with requirements. It establishes the foundation for safe and efficient operations in the oil and gas sector. Effective commissioning processes contribute to the smooth transition from construction to operations, minimizing safety risks, optimizing performance, and enhancing overall project success.

SIMOPS

SIMOPS stands for Simultaneous Operations. It refers to the practice of performing multiple activities or operations concurrently in the same workspace or facility while ensuring that they do not interfere with each other and maintain safety and productivity.

SIMOPS is commonly encountered in industries such as oil and gas, construction, manufacturing, and mining where multiple tasks need to be accomplished simultaneously to optimize project timelines and efficiency. This could involve activities such as drilling, construction, maintenance, production, and other operations taking place in the same location or vicinity.

The key considerations in SIMOPS include:

1. Risk Management: Assessing and mitigating potential risks and hazards that may arise from simultaneous operations, including the identification of potential conflicts, equipment interactions, safety concerns, and the implementation of appropriate safety measures.

2. Planning and Coordination: Develop comprehensive plans and procedures that outline the sequence, timing, and necessary

controls for each activity to ensure smooth coordination and prevent conflicts. Effective communication and collaboration among all stakeholders are critical for successful SIMOPS.

3. Permitting and Authorization: Obtaining necessary permits, clearances, and regulatory approvals specific to each operation conducted simultaneously. Adherence to safety standards, regulations, and environmental requirements is of utmost importance.

4. Training and Competency: Ensuring that personnel involved in SIMOPS are well-trained, competent, and aware of their roles and responsibilities. Proper training on safety procedures, equipment operation, emergency response, and any specialized requirements associated with specific operations is essential.

5. Monitoring and Control: Implementing robust monitoring systems to continually assess the progress, performance, and compliance of simultaneous operations. This enables prompt identification and resolution of any issues or potential conflicts that may arise during the process.

6. Emergency Preparedness: Establishing comprehensive emergency response plans that address potential emergencies or incidents that may occur during SIMOPS. This includes appropriate protocols for evacuation, emergency shutdowns, first aid, firefighting, and rescue operations.

While SIMOPS offers advantages such as time and cost optimization, it demands meticulous planning, effective coordination, and rigorous safety practices to ensure the well-being of personnel, the preservation of assets, and the smooth execution of operations. Properly managed SIMOPS can enhance operational efficiency, productivity, and project success.

COMT

COMT stands for Commissioning, Operations, Maintenance, and Training. In the EPC (Engineering, Procurement, and Construction) industry, COMT refers to a phase or stage of a project that focuses on the commissioning, operation, maintenance, and training aspects of a facility or system.

During the COMT phase, the project transitions from the construction and installation stage to a state where the facility or system is ready for operation, maintenance, and utilization. This phase involves several important activities:

1. Commissioning: This involves the testing and verification of all systems, equipment, and components to ensure proper functionality and adherence to design specifications. Commissioning activities include system testing, start-up, performance checks, and identifying and resolving any issues or deficiencies.

2. Operations: Once the facility or system has been commissioned and deemed operational, the focus shifts to its day-to-day operation. This includes managing processes, and production, monitoring performance, troubleshooting any operational challenges, and maintaining overall efficiency.

3. Maintenance: Maintenance activities are essential to ensuring the ongoing reliability, safety, and optimal performance of the facility or system. This includes preventive maintenance, inspections, repairs, periodic servicing, and scheduled downtime for necessary maintenance tasks.

4. Training: Proper training is crucial to enable personnel to effectively operate and maintain the facility or system. Training programs may be conducted to familiarize operators with equipment, processes, safety protocols, emergency procedures, and maintenance techniques. This ensures that the workforce is competent in managing the facility or system efficiently.

The COMT phase is critical for the successful transition from project completion to operational readiness. It ensures that all systems and personnel are prepared to operate, maintain, and sustain the facility or system effectively. Proper commissioning, seamless operations, proactive maintenance, and adequate training contribute to overall project success and long-term operational excellence.

First production output

In an EPC (Engineering, Procurement, and Construction) project within the oil and gas sector, "first production output" refers to the initial or inaugural production of hydrocarbons or any other commercial output from the project facilities. It signifies the successful start of production operations, typically after the commissioning phase and when the project has achieved the capability to extract, process, and deliver the intended product(s).

1. Facility Readiness: Before reaching the stage of the first production output, the project's facilities, including wells, processing units, pipelines, storage systems, and associated infrastructure, must be constructed, commissioned, and tested to ensure operational readiness.

2. Testing and Preparatory Activities: In preparation for the first production, various testing and pre-commissioning activities are conducted to verify the functionality, performance, and safety of the production systems. These activities may include equipment testing, system integration checks, calibration of instruments, operational procedures review, and necessary regulatory compliance assessments.

3. Reservoir Development and Well Completion: In the context of oil and gas projects, the development of the reservoir and the successful completion of wells are essential prerequisites for achieving the first production output. Reservoir development involves drilling and completing wells, followed by activities such as perforation, stimulation, and completion tests to establish the connectivity and productivity of the reservoir.

4. Commencement of Production Operations: Once all necessary preparations are complete, and the project has obtained the

necessary regulatory approvals and permits, production operations can commence. This includes starting equipment, initiating well production, processing and separating hydrocarbons, and transporting the produced fluids to storage or downstream facilities.

5. Monitoring and Optimization: After achieving the first production output, meticulous monitoring and optimization activities are conducted to ensure efficient operations, enhance production rates, maximize recovery, and maintain quality standards. This may involve ongoing performance testing, data analysis, adjustments in production parameters, and plan optimization to improve productivity.

6. Flow Assurance and Stabilization: During the initial production phase, flow assurance measures are implemented to mitigate and manage any potential flow-related issues, such as hydrate formation, scaling, or deposition. The emphasis is on stabilizing the process, optimizing production conditions, and ensuring the consistent and reliable delivery of output with minimal disruptions.

The achievement of the first production output marks a significant milestone in an EPC project within the oil and gas sector. It signifies the successful transition from the construction and commissioning stages to full-scale production, generating revenue and contributing to the overall success and viability of the project.

Performance Test

Performance Test Run (PTR) will be conducted to demonstrate that each unit and facility, as applicable, can achieve the Performance Guarantees set out in the CONTRACT.

Before the PTR, Operation Test Run (OTR) will be conducted, and the plant will ramp up production and establish stable production at 100% plant load.

Operation Test, unit Performance Test, and overall Performance Test will be conducted by CONTRACTOR as per the plan approved by COMPANY.

All the data verification, collection, and reporting will be done by CONTRACTOR.

Final PTR reporting shall be produced by the CONTRACTOR and approved by COMPANY.

RFSU

Ready for start-up

In the oil and gas sector, RFSU stands for Ready for Start-Up. It refers to the stage in EPC (Engineering, Procurement, and Construction) projects where the facility, equipment, and systems are deemed ready to begin operations and production.

When it comes to oil and gas EPC projects, achieving the RFSU stage is of utmost importance as it signifies that the project has progressed through various stages, such as engineering design, equipment procurement, construction, and commissioning, to reach a level of completion where it is safe and fully functional for operation.

Key factors and considerations in reaching the RFSU stage include:

1. Safety Standards: Safety is paramount in the oil and gas industry. Before reaching the RFSU stage, all necessary safety measures must be in place and rigorously tested to ensure the protection of personnel, assets, and the environment. This includes compliance with safety regulations and standards, implementation of safety systems, and thorough risk assessments.

2. Equipment and Systems Readiness: All equipment and systems involved in the production process need to undergo comprehensive testing to ensure they are functioning properly. This includes mechanical integrity testing, electrical testing, and

system integration testing. Equipment and systems readiness is crucial to guarantee smooth and efficient operations.

3. Commissioning and Performance Testing: Commissioning activities, such as functional testing, calibration, and performance testing, are conducted to ensure that all components and subsystems are operating as intended. These tests typically involve simulating real-world operating conditions to validate the performance, efficiency, and reliability of equipment and systems.

4. Documentation and Regulatory Compliance: All necessary documentation, permits, licenses, and certificates must be obtained and maintained to ensure compliance with relevant regulatory requirements. This includes obtaining environmental permits, operating licenses, and compliance with health and safety regulations.

5. Training and Competency: Before start-up, personnel involved in the operation and maintenance of the facility are trained and equipped with the necessary skills and knowledge to carry out their respective roles effectively. Training programs, such as HSE (Health, Safety, and Environment) training and operational procedures, are implemented to ensure personnel competency and regulatory compliance.

Once all these factors are taken into account, and the project has successfully met the requirements, it is considered ready for start-up (RFSU). At this stage, the facility is prepared to begin oil and gas production, adhering to all necessary safety protocols and quality standards.

The achievement of the RFSU stage in EPC projects is a significant milestone indicating that the project has been executed successfully and is prepared for safe and efficient operations in the oil and gas industry.

Operation and maintenance

Operation and maintenance in an EPC (Engineering, Procurement, and Construction) project within the oil and gas sector refers to the ongoing activities and procedures aimed at managing and optimizing the operation of the project's facilities, equipment, and systems for extended periods after the construction and commissioning phase. It involves the effective management, monitoring, and maintenance of the assets to ensure safe, reliable, and efficient operations throughout the project's lifecycle.

1. Operation: The operation phase involves the day-to-day management and control of the project facilities and processes. This includes the regular operation of the production systems, monitoring of production rates, adherence to operating procedures, and optimization of production parameters to maximize efficiency and output.

2. Maintenance: Maintenance activities aim to ensure the continued functioning and reliability of assets, equipment, and systems. It involves planned and preventive maintenance routines such as inspections, servicing, repairs, and replacement of components as necessary. This includes monitoring equipment condition, identifying and rectifying faults or failures, implementing proper lubrication, and adhering to maintenance schedules.

3. Health, Safety, and Environmental (HSE) Compliance: Operation and maintenance activities prioritize the implementation of health, safety, and environmental standards and protocols to protect personnel, assets, and the surrounding environment. This includes adhering to safety procedures, conducting regular safety inspections, maintaining emergency response plans, and monitoring and minimizing environmental impacts.

4. Asset Performance and Optimization: Operation and maintenance activities focus on enhancing asset performance and optimizing operational efficiency. This includes monitoring key performance

indicators, analyzing operational data, identifying opportunities for improvement, and implementing measures to optimize energy consumption, reduce operational costs, and increase overall asset performance.

5. Inspection and Testing: Regular inspections and testing are conducted during the operation and maintenance phase to ensure compliance with regulatory requirements and to verify the integrity and reliability of critical equipment and systems. This includes non-destructive testing (NDT), operational testing, periodic equipment inspections, integrity checks, and safety audits.

6. Training and Skill Development: Operation and maintenance activities involve providing ongoing training and skill development programs for personnel involved in operating and maintaining the project's facilities. This ensures that the workforce remains knowledgeable, competent, and up-to-date with the latest technologies, operating procedures, and safety practices relevant to the oil and gas sector.

Effective operation and maintenance practices are crucial for the long-term success, safety, and profitability of EPC projects in the oil and gas sector. By diligently managing and maintaining assets, optimizing performance, and adhering to regulatory and environmental standards, operation and maintenance activities contribute to the sustainable and efficient operation of the project throughout its lifecycle.

Identification and traceability requirement

In EPC (Engineering, Procurement, and Construction) projects in the oil and gas sector, identification and traceability requirements are essential to ensure accountability, quality control, and compliance with industry standards and regulations. These requirements help track and trace materials, equipment, and components throughout the project lifecycle. Here are key aspects of identification and traceability requirements in oil and gas EPC projects:

1. Material Identification: Each material used in the project requires a unique identification number or tag to track its origin, specifications, and compliance with industry standards. This identification can include barcode labels, RFID tags, or unique serial numbers that are linked to relevant documentation, such as material certificates, inspection reports, and test results.

2. Material Testing and Certification: Materials used in the oil and gas industry must undergo stringent testing to ensure they meet specific requirements for strength, corrosion resistance, heat resistance, and other crucial factors. The traceability requirements involve maintaining records of testing, including test reports, certificates of conformance, and material manufacturer information.

3. Equipment and Component Identification: Similar to materials, equipment and components used in EPC projects should be identified and tracked throughout their lifecycle. This includes assigning unique identification numbers, labels, or tags to equipment and components, along with associated documentation such as inspection reports, manufacturing data, and operational records.

4. Vendor Documentation and Certificates: Traceability requirements include the collection and management of all vendor-related documentation, such as material certificates, manufacturing procedures, quality control plans, and certificates of compliance. Ensuring that vendors provide comprehensive documentation and maintaining a traceability record for each vendor helps guarantee the quality and integrity of equipment and components.

5. Inspection and Testing Records: Throughout the EPC project, various inspections and tests are conducted to validate the quality, performance, and compliance of equipment, structures, and systems. Maintaining detailed records of inspection and testing

activities, including checklists, reports, and non-conformance records, supports traceability and aids in quality control and future reference.

6. Change Management: Identification and traceability involve effective change management practices. Any changes made to equipment, components, or materials are documented, including revision numbers, reasons for changes, and approval processes. This ensures accurate traceability of all project modifications and establishes a clear audit trail.

7. Document Control and Verification: Robust document control practices are crucial for maintaining traceability throughout EPC projects. Proper organization, version control, verification procedures, and access controls ensure that documents, such as specifications, procedures, inspection records, and certifications, can be easily retrieved, reviewed, and verified.

8. Regulatory Compliance: Traceability requirements also involve demonstrating compliance with industry standards, codes, and regulations. Maintaining records of compliance certificates, permits, and regulatory documentation ensures adherence to necessary safety, environmental, and operational requirements throughout the project lifecycle.

Effective implementation of identification and traceability requirements in oil and gas EPC projects provides transparency, quality control, and accountability. It enables project stakeholders to validate materials, equipment, and components, ensuring that they meet specifications, regulatory requirements, and industry standards. Ultimately, identification and traceability play a crucial role in ensuring the integrity and safety of EPC projects in the oil and gas sector.

TITBITS

Wabi Sabi:

Wabi Sabi is an aesthetic philosophy that cherishes the beauty of imperfection, impermanence, and simplicity. It finds value in objects, experiences, and moments that reflect the passage of time and the natural processes of decay and change. Wabi Sabi encourages us to appreciate the beauty in the imperfect, the unfinished, and the unconventional.

The concept of Wabi Sabi is deeply rooted in Japanese art, design, and philosophy. It reminds us to find beauty in the simplicity of nature, the weathering of materials, and the fleetingness of life itself. Rather than seeking perfection, Wabi Sabi teaches us to embrace and celebrate the unique characteristics and flaws that make things or moments truly authentic and beautiful.

Embrace imperfection: Appreciate the beauty and uniqueness in flaws, imperfections, and asymmetry.

- Cultivate simplicity: Simplify your surroundings and lifestyle, finding beauty in minimalism and uncluttered spaces.
- Focus on present-moment awareness: Practice mindfulness and fully immerse yourself in the present, savoring the transient moments and appreciating the here and now.
- Emphasize the natural and authentic: Appreciate the beauty in natural materials, organic forms, and unadorned simplicity.

TITBITS

Personal identity theft

Personal identity theft in the cyber realm, often referred to as **cyber identity theft**, is when a cybercriminal gains unauthorized access to your personal information such as your name, Social Security number, or credit card details. They may use this information to commit fraud, such as opening new accounts, making purchases, or even obtaining government benefits in your name.

To safeguard against cyber identity theft, it is crucial to adopt a proactive and vigilant approach. Here are some key strategies:

- **Use strong passwords** and change them regularly.
- Enable **multi-factor authentication** where available for an extra layer of security.
- Be cautious about the information you share on **social media** and other online platforms.
- Keep your **software and antivirus** programs up-to-date to protect against malware and hacking attempts.
- Monitor your **credit reports** and financial statements for any unauthorized activity.
- Be wary of **phishing emails** or messages that ask for personal information.
- **Shred sensitive documents** before disposal to prevent dumpster diving.
- Use **identity theft protection services** that can alert you to potential threats and help you recover if your identity is stolen.

While you cannot completely eliminate the risk of identity theft, these steps can significantly reduce your vulnerability to such threats. Stay informed about the latest security practices and be proactive in protecting your personal information.

CHAPTER 9

Project Support Functions

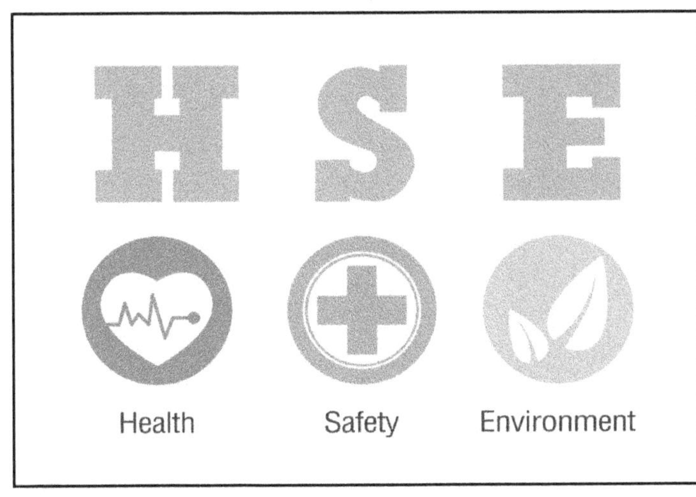

Health Safety Environment and Security (HSES)

The abbreviation changes like HSE, SHE, SHES, etc. depending on the organization. Don't worry about it.

HSE is a common department across any project. No project can start its work without HSES in place.

The objective of HSE is to ensure that the project is completed SAFELY, (without any major incidents), Employee health is taken care and finally, the environment is also protected and not spoilt. There are respective regulatory requirements applicable for HSES, depending on the country and the law of the land.

Depending on the country, the HSE standards and requirements may vary. Countries like Australia, the UK, etc., have very stringent HSE requirements and standards to be implemented.

HSES documentation list:

1. HSES policy for the organization
2. HSES manual
3. HSES procedures (SOP)
4. HSES inspections and walk-throughs
5. HSES audits
6. HSES campaigns
7. HSES records

There are a few ISO standards like ISO 14001, and OHSAS 45001, stipulate the requirements for HSES requirements.

Over and above these, the local legal requirements play a vital role.

HSES statistics reflect the culture of a company.

Environment

Environmental Impact Assessment (EIA) is a crucial process in the oil and gas sector for evaluating and mitigating the potential environmental impacts associated with exploration, production, transportation, and refining activities. It provides a comprehensive analysis of the potential environmental consequences of proposed projects, ensuring adherence to regulatory requirements and environmental sustainability.

The EIA process typically involves the following steps:

1. Screening: This initial step determines whether a project requires a full-scale EIA. It evaluates the project's size, location, and potential impacts to identify if significant environmental effects are likely.

2. Scoping: Scoping involves identifying the potential environmental impacts that need to be assessed during the EIA process. Stakeholder consultations are conducted to gather input, concerns, and information for an inclusive assessment.

3. Baseline Data Collection: Environmental baseline studies are conducted to document the current state of the environment in the project area. This includes assessing air and water quality, biodiversity, land use, socio-economic factors, and cultural heritage.

4. Impact Prediction: The EIA evaluates potential impacts on various environmental components, such as air, water, soil, flora, fauna, and socio-economic factors. Predictive models, environmental data, and expert judgment are used to estimate the magnitude and significance of potential impacts.

5. Impact Mitigation: Mitigation measures are identified and developed to minimize or eliminate potential environmental impacts. This may involve designing and implementing pollution control measures, habitat restoration programs, waste management strategies, and community engagement initiatives.

6. Impact Assessment and Reporting: The environmental impacts identified, and their mitigation measures are compiled into a comprehensive report. This report assesses the implications of

the project on the environment, analyzes the effectiveness of proposed mitigation measures, and suggests alternatives for reducing adverse impacts.

7. Decision-Making: The EIA report is reviewed by regulatory authorities, who make informed decisions based on the findings and recommendations. The decision may involve project approval, denial, or conditional approval, taking into account the environmental, social, and economic aspects of the proposed project.

8. Monitoring and Compliance: Post-approval, monitoring programs are implemented to ensure compliance with mitigation measures and to assess the actual environmental impacts of the project. This facilitates continuous improvement, adaptive management, and accountability.

The EIA process in the oil and gas sector aims to achieve sustainable development by balancing economic growth with environmental protection. It provides decision-makers, stakeholders, and the public with valuable information to make informed choices, fostering responsible resource development and protecting ecological integrity.

Quality

In the oil and gas sector, where projects entail complex engineering, vast investments, and potentially hazardous operations, ensuring quality in project execution is of utmost importance. Quality encompasses adherence to specifications, industry standards, and regulations, as well as the ability to deliver projects that meet or exceed client expectations. This section will delve into the reasons why maintaining quality holds significant value in the execution of projects within the oil and gas sector.

1. Safety:

Safety is paramount in the oil and gas industry. High-quality project execution incorporates robust safety measures and adherence to industry standards, reducing the risk of accidents, injuries, and environmental incidents. Quality-driven project execution entails detailed risk assessments, proper equipment testing, adherence to safety codes, and rigorous inspections, all of which contribute to safeguarding personnel, assets, and the surrounding environment.

2. Reliability and Efficiency:

Quality project execution ensures the reliability and efficiency of systems, equipment, and processes. It involves meticulous design, engineering, and fabrication practices, as well as the use of reliable materials and components. Implementing quality-driven techniques enhances the longevity and performance of assets, preventing costly downtime, operational disruptions, and equipment failures. Moreover, quality execution optimizes processes, minimizing waste, and ensuring efficient utilization of resources, thereby reducing project delays and cost overruns.

3. Compliance and Regulatory Requirements:

The oil and gas sector operates under various legal and regulatory frameworks established to safeguard public health, the environment, and industry interdependencies. High-quality project execution ensures compliance with these regulations, industry standards, and permits. Thorough quality control and inspection processes verify that projects meet the specified requirements, ultimately building trust and demonstrating accountability to regulatory bodies, stakeholders, and the community.

4. Client Satisfaction and Reputation:

Quality execution directly impacts customer satisfaction, which is vital for the long-term success of projects within the oil and gas sector. Delivering a high-quality project that meets or exceeds client expectations ensures their confidence, trust, and ongoing collaboration. It builds a positive reputation for contractors and stakeholders, leading to valuable partnerships, repeat business, and referrals. Enhanced client satisfaction strengthens the position of a company in the competitive market, contributing to its long-term growth and success.

5. Risk Mitigation and Cost Control:

Implementing quality-driven project execution practices helps in mitigating risks and controlling costs. By identifying and rectifying potential issues early in the project lifecycle, costly rework, delays, and budget overruns can be avoided. Efficient quality control processes ensure that identified risks are addressed through preventive measures, avoiding potential setbacks and associated expenses.

6. Compliance with International Standards:

Projects in the oil and gas sector are often executed amidst a global context. Maintaining high-quality standards in project execution ensures compliance with international norms, codes, and best practices. Adhering to internationally recognized quality standards enhances the reputation and competitiveness of companies operating in the global market, enabling access to wider opportunities and demonstrating a commitment to excellence.

Quality plays a crucial role in project execution within the oil and gas sector. From ensuring safety and reliability to complying with regulations and satisfying clients, maintaining high-quality standards leads to successful project outcomes. Quality-driven approaches minimize risks, control costs, and enhance the sector's overall reputation for delivering projects efficiently and responsibly. Embracing quality as an intrinsic value throughout the project lifecycle ultimately leads to increased productivity, improved safety performance, and sustainable growth in the oil and gas industry.

The key difference between QA and QC

Criteria	Quality Assurance (QA)	Quality Control (QC)
Scope	Preventive	Reactive
Timing	Implemented throughout the project lifecycle	During project execution
Objective	Ensure the right processes and systems are in place to achieve desired quality outcomes	Verify that products or deliverables meet specified quality standards
Focus	Prevention of defects and non-conformities	Detection and correction of defects or non-conformities
Activities	Quality management system establishment, audits, continuous improvement	Inspections, testing, monitoring, sampling, corrective actions
Purpose	Establish and maintain quality standards, processes, and systems	Verify and validate products or deliverables against established requirements
Outcome	Consistent quality outcomes and continuous improvement	Compliance with quality standards and resolution of non-conformities

Quality Assurance

Definition: QA is a set of activities designed to ensure that quality requirements are met before a product or service is delivered to the customer. QC, on the other hand, is a set of activities designed to ensure that the product or service meets the desired level of quality after it has been produced.

Focus: QA focuses on preventing defects from occurring by ensuring that processes are in place to produce a quality product or service. QC focuses on identifying and correcting defects that have already occurred.

Responsibility: QA is the responsibility of the entire organization, while QC is typically the responsibility of a dedicated quality control department or team.

Timeframe: QA is a proactive process that occurs throughout the entire development or production process. QC is a reactive process that occurs after production is completed.

Techniques: QA involves techniques such as process improvement, statistical process control, and quality planning. QC involves techniques such as inspection, testing, and quality control charts.

In summary, QA and QC are both important aspects of quality management, but they focus on different stages of the quality management process. QA is focused on preventing defects from occurring, while QC is focused on identifying and correcting defects that have already occurred

Key activities of QA in a typical project are:

- Develop the Project Quality plan. (Ref: ISO 10005:2018)
- Prepare System Audit procedure and schedule
- Monitor and measure the quality performance
- Conduct trainings and toolbox talks to create quality awareness
- Conduct Management review meetings
- Control and handle nonconformances
- Perform root cause analysis
- Interact with client QA personnel.

The QA department develops the project quality plan, which is one of the key documents in any project to ensure quality.

The other key procedures are.

- Project execution plan
- Project control plan
- Project interface management plan
- Project document management plan
- Project IT plan
- Project communication and coordination plan
- Project procurement plan
- Project Construction and Commissioning plan

QA in project management

Quality assurance (QA) is a crucial aspect of project management, ensuring excellence, customer satisfaction, and adherence to standards. To achieve these objectives, the implementation of appropriate QA tools is essential. These tools allow for efficient planning, execution, monitoring, and control of quality-related activities throughout the project lifecycle. This section explores several key QA tools that contribute to enhancing project quality and driving successful outcomes.

1. Checklists:

Checklists serve as handy tools for ensuring that all necessary tasks and quality requirements are addressed. They provide a systematic approach to verifying compliance with standards, procedures, and contract obligations. By following a checklist, project teams can effectively monitor and track quality-related activities, enhancing consistency and diligence in project execution.

2. Audit Systems:

Audit systems provide a robust framework for assessing project performance, compliance, and adherence to quality standards.

Conducting regular audits enables the identification of potential shortcomings, non-compliance, and areas for improvement. Through the systematic review of processes, procedures, and records, audit systems help maintain a high level of quality, enabling early detection and resolution of issues.

3. Statistical Process Control (SPC):

SPC is a statistical tool used to analyze data and monitor process variations. By collecting and analyzing data over time, SPC helps identify trends, patterns, and deviations from established quality baselines. This tool enables project teams to make data-driven decisions, identify potential issues before they escalate, and implement timely corrective actions to maintain process control and achieve consistent quality outcomes.

4. Root Cause Analysis (RCA):

RCA is a problem-solving technique used to identify the underlying causes of quality issues or incidents. It involves a systematic investigation that aims to uncover the root causes rather than solely addressing superficial symptoms. By determining the root cause of a problem, project teams can implement targeted corrective actions, preventing the recurrence of similar issues in the future and continuously improving the overall quality.

5. Risk Assessment Tools:

Risk assessment tools assist in identifying potential risks and their impact on project quality. These tools help project teams analyze and prioritize risks, developing proactive mitigation measures and contingency plans. By incorporating risk assessment tools into the QA process, project teams can anticipate and address potential quality-related risks, ensuring that the project stays on track and meets or exceeds quality expectations.

Project Quality Plan (PQP) (ISO 10005:2018)

Project Quality Plan in the Oil and Gas Industry EPC Projects: Ensuring Quality Excellence across Phases

The project quality plan is a crucial component of EPC (Engineering, Procurement, and Construction) projects in the oil and gas industry. It establishes a comprehensive framework, guidelines, and procedures to ensure quality excellence throughout all project phases. Here's an overview of a project quality plan that covers the key phases of an EPC project in the oil and gas industry:

1. Design Phase:

During the design phase, the project quality plan focuses on establishing a robust design management process. It includes procedures for quality control and assurance, design reviews, verification, and validation activities. The plan ensures compliance with design codes, standards, and project specifications while promoting constructability, safety, and reliability.

2. Procurement Phase:

In the procurement phase, the project quality plan outlines procedures for vendor qualification, selection, and evaluation. It details quality requirements for procured materials, equipment, and services. The plan includes controls for incoming inspection, supplier audits, and quality assurance during the procurement process, ensuring adherence to specified standards and project requirements.

3. Construction Phase:

In the construction phase, the project quality plan focuses on implementing quality assurance and control measures for construction activities. It covers procedures for inspection and testing, welder qualification, non-destructive testing, and compliance with applicable codes and specifications. The plan emphasizes safety protocols, quality documentation, and project-specific quality objectives.

4. Commissioning and Start-up Phase:

During the commissioning and start-up phase, the project quality plan includes procedures for pre-commissioning checks, functional testing, performance verification, and system integration. It ensures the proper execution of documented commissioning plans, addressing equipment, control systems, safety systems, and operational readiness. The plan emphasizes the importance of thorough documentation, testing, and handover processes.

5. Operation and Maintenance Phase:

In the operation and maintenance phase, the project quality plan outlines procedures for ongoing quality management, inspection, and maintenance activities. It covers preventive and corrective maintenance, reliability-centered maintenance practices, and continuous improvement initiatives. The plan emphasizes the importance of performance monitoring, data analysis, and feedback loops for enhancing operational efficiency and reliability.

Throughout all phases, the project quality plan incorporates procedures for quality audits, management reviews, and risk management activities. It emphasizes a proactive approach to quality management, promoting adherence to industry standards, regulatory requirements, and client specifications. The plan fosters a culture of continuous improvement, with mechanisms for lessons learned, root cause analysis, and corrective actions. Detailed documentation, training programs, and communication channels are delineated to ensure awareness, understanding, and effective implementation of quality processes.

By having a comprehensive project quality plan in place, EPC projects in the oil and gas industry can consistently deliver high-quality outcomes, meeting client expectations, and industry standards, and ensuring long-term operational excellence and safety.

Contents of a Project Quality Plan

1. Introduction
   1.1 Purpose of the Quality Plan
   1.2 Project Overview
   1.3 Definitions and Acronyms

2. Quality Management System
   2.1 Quality Policy
   2.2 Roles and Responsibilities
   2.3 Document Control Procedures
   2.4 Change Control Procedures

3. Quality Planning
   3.1 Project Quality Objectives
   3.2 Quality Assurance and Control Activities
   3.3 Quality Management Approaches
   3.4 Project Quality Milestones and Deliverables

4. Design Phase Quality Activities
   4.1 Design Management Procedures
   4.2 Design Verification and Validation
   4.3 Design Reviews and Sign-Off
   4.4 Design Documentation Control

5. Procurement Phase Quality Activities
   5.1 Vendor Qualification and Selection
   5.2 Procurement Quality Requirements
   5.3 Incoming Inspection Procedures
   5.4 Supplier Audits and Performance Monitoring

6. Construction Phase Quality Activities
   6.1 Quality Control Plans and Procedures
   6.2 Inspection and Testing Activities
   6.3 Non-Destructive Testing Requirements
   6.4 Compliance with Codes and Specifications

7. Commissioning and Start-up Phase Quality Activities
   7.1 Commissioning Plans and Procedures
   7.2 Pre-commissioning Checks and Testing
   7.3 Performance Verification and System Integration
   7.4 Documentation and Handover Processes

8. Operation and Maintenance Phase Quality Activities
   8.1 Preventive and Corrective Maintenance Procedures
   8.2 Reliability-centered Maintenance Practices
   8.3 Performance Monitoring and Data Analysis
   8.4 Continuous Improvement Initiatives

9. Quality Audits and Reviews
   9.1 Internal and External Audits
   9.2 Management Reviews and Reporting
   9.3 Inspection and Testing Audits
   9.4 Corrective and Preventive Actions

10. Documentation and Records Management
    10.1 Document Control Procedures
    10.2 Records Management and Retention

11. Training and Competency Development
    11.1 Training Needs Assessment
    11.2 Training Programs and Plans
    11.3 Competency Evaluation and Improvement

12. Communication and Stakeholder Engagement
    12.1 Communication Channels and Protocols
    12.2 Stakeholder Engagement and Feedback
    12.3 Lessons Learned and Knowledge Sharing

Appendices:
   A. List of Acronyms and Definitions
   B. Project Quality Forms and Templates
   C. References and Standards
   D. Glossary of Terms

Note: The table of contents can be adjusted based on the specific requirements and scope of the project.

QUALITY MANAGEMENT SYSTEMS – International standards

ISO standards for oil and gas projects cover various aspects of quality, environmental management, occupational health and safety, risk management, and energy management. Here is a quick comparison of some key ISO standards relevant to the oil and gas sector:

1. ISO 9001: Quality Management System (QMS)

- Focuses on ensuring quality in all aspects of project delivery, from planning to execution and continuous improvement.
- Enhances customer satisfaction, increases efficiency, and minimizes errors and rework.
- Applicable to organizations involved in oil and gas exploration, production, and service delivery.

2. ISO 14001: Environmental Management System (EMS)

- Provides a framework for organizations to identify and manage environmental impacts associated with oil and gas projects.

- Aim to minimize negative environmental effects, comply with regulations, and improve sustainability performance.
- Includes aspects such as waste management, pollution prevention, and resource conservation.

3. ISO 45001: Occupational Health and Safety Management System (OHSMS)

- Focuses on ensuring a safe and healthy work environment for employees and contractors involved in oil and gas projects.
- Identifies and prevents workplace hazards, reduces accidents, and improves overall occupational health and safety performance.
- Helps organizations demonstrate a strong commitment to employee well-being and comply with legal requirements.

4. ISO 31000: Risk Management

- Provides principles, framework, and guidelines for effective risk management in oil and gas projects.
- Helps organizations identify, assess, and mitigate risks, ensuring better decision-making and optimal project outcomes.
- Emphasizes the importance of a proactive and systematic approach to risk management.

5. ISO 50001: Energy Management System (EnMS)

- Designed to assist organizations in improving energy efficiency and reducing energy consumption.
- Suitable for oil and gas projects where energy-intensive operations are involved.
- Helps identify energy-saving opportunities, adopt energy-efficient practices, and meet energy-related legal and regulatory requirements.

It is important to note that these ISO standards are complementary and can be integrated to form a comprehensive management system. Organizations in the oil and gas sector can align their processes with multiple ISO standards to enhance overall performance, streamline operations, mitigate risks, and demonstrate a commitment to quality, environmental sustainability, health and safety, and energy management.

QMS hierarchy of documentation

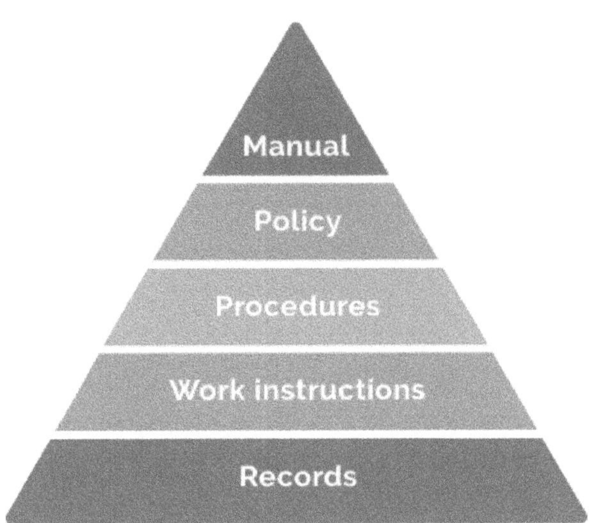

Quality management System audit process

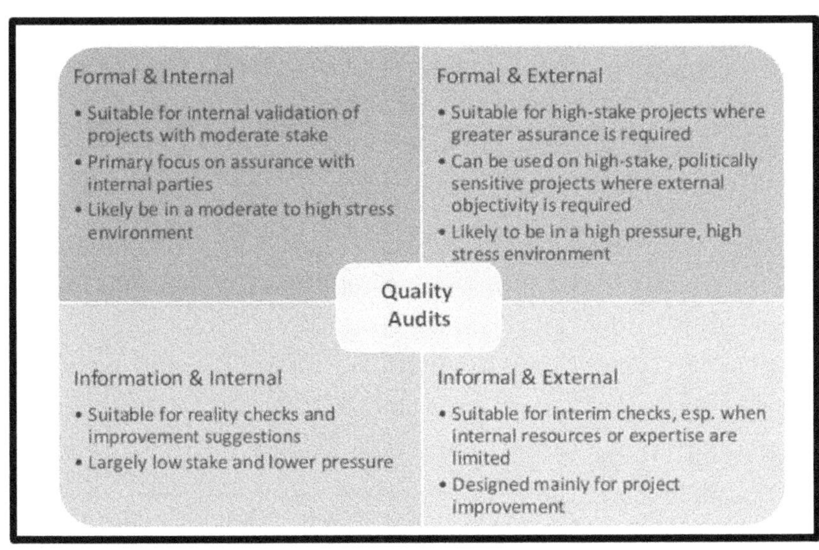

A QMS (Quality Management System) audit is conducted to assess the effectiveness and compliance of a company's quality management system. The procedure typically involves the following steps:

1. Planning: The auditor establishes the objectives, scope, and criteria for the audit. They identify the areas to be audited and plan the activities accordingly.
2. Opening meeting: The auditor holds a meeting with the relevant personnel to explain the purpose of the audit, the audit process, and the expectations from the auditee.
3. Document review: The auditor examines the company's quality manual, procedures, work instructions, records, and other relevant documents to ensure compliance with the defined QMS requirements.
4. Fieldwork: The auditor conducts interviews and observations to gather information and evidence related to the implementation of the QMS. They may interact with employees at different levels and review specific processes or functions.
5. Data analysis: The auditor analyzes the collected evidence and compares it against the defined criteria to identify any deviations, non-compliance, or areas for improvement.
6. Audit findings: The auditor documents any findings, including observations, non-conformities, and potential opportunities for improvement. They classify the findings based on their severity or significance.
7. Closing meeting: The auditor holds a meeting with the auditee to present the audit findings. This allows for discussion, clarification, and agreement on any identified non-conformities and their resolution.
8. Corrective actions: The auditee develops and implements corrective actions to address the identified non-conformities and areas for improvement. These actions aim to prevent recurrence and enhance the effectiveness of the QMS.

9. Follow-up: The auditor may conduct a follow-up audit to verify the implementation and effectiveness of the corrective actions taken by the auditee.

10. Audit report: The auditor prepares a comprehensive audit report summarizing the audit objectives, scope, findings, and conclusions. The report may include recommendations for improvement and suggestions for further actions.

11. Review and improvement: The auditee reviews the audit report, learns from the findings, and takes appropriate actions for continuous improvement of their QMS.

It is important to note that the specific steps and procedures may vary depending on the organization's specific QMS requirements and the standards against which the audit is conducted.

The results of audits are one of the inputs for management review, training, and improvement.

Root cause analysis

Root cause analysis (RCA) is a systematic approach used to identify the underlying causes of any non-conformance related to quality in the oil and gas sector. The process typically involves the following steps:

1. Define the problem: Identify and define the non-conformance or quality issue that needs to be investigated. This could include defects, deviations from standards, equipment failures, process inefficiencies, safety incidents, or customer complaints.

2. Collect data: Gather relevant data and information related to the non-conformance. This may include inspection reports, test results, maintenance records, operational data, incident reports, and any other available documentation.

3. Apply problem-solving techniques: Use various problem-solving techniques, such as the fishbone diagram (Ishikawa diagram), 5 Whys, or Pareto analysis, to determine the possible causes of the non-conformance. These techniques help identify potential

contributing factors and direct the investigation in the right direction.

4. Analysis: Analyze the collected data, facts, and observations to identify potential root causes. This may involve conducting interviews, reviewing procedures, analyzing equipment performance, or studying process flows.

5. Validate causes: Validate the potential root causes by applying critical thinking and using additional evidence. This could involve further data analysis, conducting experiments, or seeking expert opinions.

6. Identify root cause(s): Once the potential causes have been validated, determine the primary root cause(s) of the non-conformance. This step involves identifying the most significant cause(s) that, if addressed, will prevent the non-conformance from recurring.

7. Develop corrective actions: Based on the identified root causes, develop specific and practical corrective actions to address each root cause. These actions should aim to eliminate or mitigate the root cause and prevent future non-conformances.

8. Implement and monitor: Implement the corrective actions and monitor their effectiveness. This may involve updating procedures, providing training, modifying equipment, or implementing process changes.

9. Verify effectiveness: Verify the effectiveness of the corrective actions through follow-up inspections, testing, or monitoring. This step ensures that the identified root cause(s) have been successfully addressed and that the non-conformance has been resolved.

10. Document and communicate: Document the entire RCA process, including the identified root causes and the implemented corrective actions. Communicate the findings and actions to the relevant stakeholders, ensuring transparency and knowledge sharing.

By systematically performing root cause analysis in the oil and gas sector, organizations can identify and address the underlying causes of non-conformances and improve the overall quality and performance of their operations.

Lessons learned

The lessons learned collation once the project is completed is an important activity by the project management team.

The QA personnel will coordinate this activity and maintain them in a database.

when a new project is kicked off, Lessons are picked up from this database and lessons learned workshop is conducted for the new project.

Preventing past mistakes are repeated in the current projects is the main objective of lessons.

Generally, the project engineer will coordinate with the concerned lead engineers for the implementation of lessons learned.

Unfortunately, many organizations do not give proper importance to lessons learned, rather consider them as a ritual.

Japanese Quality Techniques

Japanese quality techniques, forged through the philosophy of Kaizen (continuous improvement) and a deep commitment to excellence, have had a profound impact on global quality management practices. This section explores some of the key Japanese quality techniques that have revolutionized industries worldwide, driving higher product and service standards, improving efficiency, and fostering a culture of relentless improvement.

1. Total Quality Management (TQM):

Total Quality Management is a Japanese quality approach that emphasizes the involvement of all members of an organization in continuously improving quality at every level. It emphasizes customer focus, process improvement, and employee empowerment. TQM promotes a holistic view of quality and places a strong emphasis on the elimination of waste and reducing variations.

2. Lean Manufacturing and the Toyota Production System (TPS):

Lean manufacturing, influenced by the Toyota Production System, focuses on streamlining operations, eliminating non-value-added activities, and optimizing efficiency. By reducing waste, improving flow, and empowering employees to identify and solve problems, lean manufacturing has become synonymous with improved quality, reduced costs, and increased productivity in diverse industries globally.

3. Just-in-Time (JIT):

Just-in-Time, an integral part of the Toyota Production System, revolutionized inventory management by emphasizing the timely delivery of materials and components. JIT aims to eliminate waste caused by overproduction, waiting time, excess inventory, and inefficient processes. By closely aligning supply with demand, JIT enables smoother operations, reduces costs, and enhances responsiveness to customer needs.

4. Kanban:

Kanban, meaning "visual card" in Japanese, is a system for visualizing and managing workflow. It uses visual cues, such as cards or boards, to represent tasks and guide the movement of work through different stages. Kanban promotes transparency, enhances communication, and facilitates efficient collaboration between team members, ultimately improving productivity and quality.

5. Poka-Yoke:

Poka-Yoke, which translates to "mistake-proofing," aims to prevent errors and defects from occurring at the source. It involves designing processes, equipment, or systems in a way that makes errors or mistakes impossible or unlikely to happen. Poka-Yoke techniques include using sensors, fixtures, and visual cues to provide immediate feedback and prevent or detect errors before they escalate.

6. 5S/5C Concepts:

The 5S/5C Concepts are principles used in workplace organizations to enhance efficiency, safety, and productivity. Each "S" or "C" represents a Japanese word that corresponds to a specific step in the process:

- Seiri (Sort): Remove unnecessary items from the workspace to reduce clutter and improve organization.
- Seiton (Set in Order): Arrange and organize necessary items logically and efficiently, making them easily accessible.
- Seiso (Shine): Clean and maintain the workspace consistently to ensure a safe and conducive environment for work.
- Seiketsu (Standardize): Establish standardized procedures and practices to sustain the improvements made in the first three steps.
- Shitsuke (Sustain): Cultivate a disciplined culture of maintaining the 5S/5C practices, continuously monitoring and improving to ensure long-term benefits.

7. Kaizen:

Kaizen embraces the philosophy of continuous improvement, encouraging small incremental changes to improve processes, products, and services over time. It involves the active participation of employees at all levels to identify and implement improvements. Kaizen fosters a culture of collaboration, creativity, and continual learning, leading to sustained enhancements in quality and operational effectiveness.

Implementing these Japanese quality techniques enables organizations to unlock higher levels of quality, productivity, and employee engagement, ultimately leading to long-term success and competitive advantage.

Emerging Trends in Quality Assurance

In the oil and gas industry, there are several emerging trends in Quality Assurance (QA) that organizations are adopting to enhance their operations and ensure compliance with industry standards. Some of the latest trends in QA about the oil and gas industry include:

1. Digital Transformation: The adoption of digital technologies such as artificial intelligence, machine learning, and automation is revolutionizing QA processes in the oil and gas industry. These technologies enable real-time data collection, analysis, and predictive maintenance, resulting in improved quality and reliability while minimizing costly downtime.

2. Risk-Based Approach: Organizations are shifting towards a risk-based approach to QA, where resources and efforts are directed towards areas of higher risk or criticality. This approach allows for more efficient allocation of resources, better identification of potential issues, and enhanced decision-making processes.

3. Integrated Quality Management Systems: The integration of Quality Management Systems (QMS) with other systems such as Health, Safety, and Environment (HSE) is gaining momentum. This integrated approach ensures seamless communication and collaboration between different functions within an organization, enhancing overall operational efficiency and reducing duplication of efforts.

4. Supplier Quality Management: With the increasing complexity of supply chains in the oil and gas industry, effective management of suppliers is crucial for maintaining high-quality standards. Organizations are implementing robust supplier quality management systems, including supplier qualification, performance evaluations, and continuous monitoring to ensure the quality of incoming materials and services.

5. Data Analytics and Advanced Analytics: The utilization of data analytics and advanced analytics techniques allows organizations to extract valuable insights from large volumes of data. By

analyzing historical and real-time data, organizations can identify trends, anomalies, and patterns, making data-driven decisions to improve quality performance and optimize operations.

6. Asset Integrity Management: In the oil and gas industry, ensuring the integrity of assets is paramount for safety and operational excellence. QA processes are evolving to include more proactive approaches to asset integrity management, such as predictive maintenance, inspection, and testing techniques, to minimize the risk of equipment failure and unplanned downtime.

7. Continuous Improvement and Lean Principles: Organizations are increasingly adopting a culture of continuous improvement and implementing Lean principles in QA processes. This involves eliminating waste, streamlining processes, empowering employees, and optimizing workflows to enhance efficiency, quality, and customer satisfaction.

By embracing these latest trends in QA, the oil and gas industry can enhance its overall quality performance, optimize operations, and ensure compliance with regulations and industry standards.

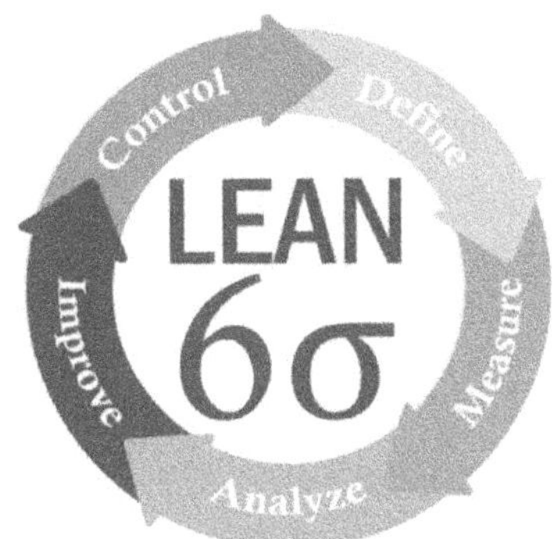

Lean Six Sigma

In today's fiercely competitive business landscape, organizations are constantly seeking ways to optimize their operations and drive continuous improvement. Among the various methodologies available, Lean Six Sigma stands out as a powerful approach that brings significant benefits to organizations across industries. This article aims to provide a professional insight into Lean Six Sigma, exploring its key principles, including the acclaimed DMAIC framework.

Lean Six Sigma combines two well-established methodologies: Lean and Six Sigma. Lean focuses on eliminating waste and increasing efficiency, while Six Sigma targets reducing defects and variations. Together, they form a robust methodology that enhances productivity, quality, and customer satisfaction.

At the core of Lean Six Sigma lies the DMAIC framework, an acronym that represents the structured approach to problem-solving and process improvement:

1. Define: In this initial phase, the focus is on clearly articulating the problem or opportunity for improvement. By defining measurable goals, understanding customer requirements, and establishing a project scope, organizations can lay a solid foundation for success.

2. Measure: In the measurement phase, data is gathered to gain a comprehensive understanding of the current state of the process. Key performance indicators (KPIs) are identified and measured, allowing organizations to assess the magnitude of the problem and establish a baseline for future improvements.

3. Analyze: The analysis phase involves a thorough examination of the data collected in the previous phase. Statistical tools and techniques, such as root cause analysis and process mapping, are employed to identify the underlying causes of issues and inefficiencies. This critical analysis helps organizations gain valuable insights and make informed decisions.

4. Improve: Once the root causes are identified, the improvement phase comes into play. Organizations brainstorm innovative solutions to address the identified issues. Through rigorous experimentation, pilot testing, and the application of Lean tools

such as value stream mapping and 5S, lasting improvements are implemented.

5. Control: The final phase of DMAIC focuses on putting control mechanisms in place to sustain the improvements achieved. Organizations establish well-defined processes, create standard operating procedures, and implement monitoring systems to prevent regression and ensure ongoing success.

As organizations dive into the world of Lean Six Sigma, a variety of jargon often surfaces. Terms such as Kaizen events (focused improvement workshops), Pareto analysis (prioritizing root causes), and Fishbone diagrams (visualizing cause and effect relationships) become part of the discourse. These specialized terms represent powerful tools and concepts that facilitate a structured approach to process improvement and optimization.

By applying Lean Six Sigma methodologies, organizations can streamline their processes, reduce waste, and enhance overall productivity. The ultimate goal is to provide higher-quality products or services, exceed customer expectations, and gain a competitive edge in the market.

While Lean Six Sigma has its roots in the corporate world, its principles can extend beyond business contexts. Applicable in various fields, it can bring about remarkable improvements in personal projects, time management, and daily routines. By embracing the problem-solving mindset and utilizing Lean Six Sigma tools, we can lead more efficient and effective lives.

Lean Six Sigma offers organizations a systematic and data-driven approach to achieving operational excellence. Through the DMAIC framework and associated tools, they can address process inefficiencies, reduce defects, and drive continuous improvement. By applying Lean Six Sigma principles in their personal lives, individuals can optimize their endeavors, minimize waste, and foster success. Lean Six Sigma is a valuable methodology that empowers individuals and organizations to unlock their true potential and achieve excellence in every aspect of their lives.

Quality control

Quality control plays a crucial role in the oil and gas industry, particularly when it comes to fabrication and construction activities. The fabrication and construction stage of oil and gas projects involves intricate engineering, stringent safety measures, and the assembly of complex systems. This section explores the importance of quality control in fabrication and construction within the oil and gas industry and examines key practices and challenges in maintaining quality throughout the process.

Importance of Fabrication and Construction Quality Control:

Fabrication and construction activities in the oil and gas industry require adherence to stringent quality standards to ensure the safe and efficient operation of assets. Quality control ensures that materials, equipment, and structures are constructed to meet specified requirements, regulations, and industry best practices. It involves rigorous inspections, testing, and verification to identify and rectify any deviations, preventing potential failures and minimizing risks to personnel, the environment, and assets. Effective quality control measures also contribute to the overall project success by reducing rework, enhancing operational efficiency, and safeguarding project schedules and budgets.

Fabrication Quality Control Practices:

1. Material Inspection and Testing:

Quality control procedures begin with rigorous inspection and testing of raw materials, components, and equipment. This includes verifying the properties and specifications of materials, conducting non-destructive testing, and ensuring compliance with industry codes and standards. Inspection processes may involve visual inspections, dimensional checks, ultrasonic testing, radiographic examinations, and other testing methods to confirm material integrity and suitability for the intended application.

2. Welding and Fabrication Inspections:

Fabrication quality control places significant emphasis on welding and fabrication processes. Qualified inspectors oversee weld procedure qualifications, welder qualifications, and welding inspections to ensure adherence to approved procedures, joint configurations, and specific fabrication specifications. Inspections focus on parameters such as weld dimensions, weld penetration, proper materials usage, and compliance with applicable welding codes and standards.

3. Documentation and Traceability:

Thorough documentation and traceability are vital throughout the fabrication process. Quality control personnel maintain comprehensive records of inspections, testing outcomes, and certifications, including material and weld traceability records. This documentation ensures that materials and components can be traced back to their source, facilitating quality assurance and providing a clear audit trail.

Construction Quality Control Practices:

1. Construction Inspections:

Construction quality control involves stringent inspections of installations, structural components, and systems. These inspections verify compliance with design specifications, engineering drawings, and safety requirements. Inspectors assess the adequacy and quality of concrete works, structural steel installations, equipment, pipelines, electrical systems, and other construction elements.

2. Installation and Assembly Checks:

Quality control personnel closely monitor the installation and assembly of equipment and systems to ensure proper alignment, calibration, and functional performance. They verify correct installation techniques, and proper use of fasteners and gaskets, and ensure compliance with manufacturer guidelines and project-specific requirements. Regular inspections during the assembly process mitigate potential issues and confirm adherence to quality standards.

3. System Testing and Commissioning:

Quality control extends to system testing and commissioning to verify proper functionality, performance, and safety. This involves conducting pressure tests, leak tests, functional tests, and the commissioning of equipment and systems. Quality control personnel work closely with testing and commissioning teams to ensure that all procedures are executed accurately, deviations are identified, and appropriate corrective actions are taken.

Challenges in Fabrication and Construction Quality Control:

The oil and gas industry faces unique challenges in maintaining effective quality control during fabrication and construction, such as ensuring compliance with diverse international standards, managing complex supply chains, and addressing potential workforce and cultural variations. Additionally, remote and harsh operating environments, stringent project timelines, and cost pressures add complexity to quality control efforts. Overcoming these challenges requires robust quality control processes, effective communication, collaboration between stakeholders, and continuous improvement practices.

Fabrication and construction quality control are critical components of successful oil and gas projects. Rigorous adherence to quality standards throughout the fabrication and construction stages ensures the safe, efficient, and reliable operation of assets. By implementing comprehensive inspection processes, testing procedures, documentation practices, and adherence to industry regulations and standards, organizations can mitigate risks, minimize defects, and optimize project outcomes.

Effective quality control practices contribute to the long-term success of oil and gas projects by enhancing safety, sustainability, and operational excellence.

Welding

Welding processes

Different welding processes and methods are used in the field of welding to join materials together. The American Welding Society (AWS) provides guidelines and standards for these various techniques. Here's an overview of some commonly used welding processes and methods as per AWS:

1. Shielded Metal Arc Welding (SMAW):

- SMAW, also known as stick welding, involves the use of a consumable electrode covered with a flux coating.
- The electrode produces an electric arc that melts the base metal, the flux coating creates a protective shield, and the electrode material acts as filler.

- SMAW is versatile, cost-effective, and can be applied to a variety of materials and thicknesses, making it widely used in construction, maintenance, and repair work.

2. Gas Metal Arc Welding (GMAW):

- GMAW, also known as MIG (Metal Inert Gas) welding, utilizes a consumable electrode wire and an inert or active gas as a shielding medium.
- The electric arc melts the electrode wire and the base metal, while the shielding gas protects the weld zone from atmospheric contamination.
- GMAW is known for its high productivity, ease of automation, and applicability to various materials, making it a preferred choice for manufacturing, fabrication, and automotive industries.

3. Flux-Cored Arc Welding (FCAW):

- FCAW is similar to GMAW, but instead of using a solid wire electrode, it employs a tubular wire electrode filled with various flux materials.
- The flux core provides shielding and also contains deoxidizers and alloying elements, offering versatility and improved control over the weld characteristics.
- FCAW is commonly used in construction, shipbuilding, and heavy equipment fabrication, as it can handle thicker materials and work well in outdoor or windy conditions.

4. Gas Tungsten Arc Welding (GTAW):

- GTAW, also known as TIG (Tungsten Inert Gas) welding, utilizes a non-consumable tungsten electrode and an inert gas for shielding.
- The electric arc forms between the tungsten electrode and the workpiece, melting the base metal while the inert gas provides a protective atmosphere.

- GTAW offers excellent control, and high-quality welds, and is suited for thinner materials, stainless steel, and non-ferrous metals, making it prevalent in aerospace, automotive, and high-purity applications.

5. Submerged Arc Welding (SAW):

- SAW involves the formation of an arc between a continuously fed wire electrode and the workpiece, while a granular flux blanket covers the weld zone.
- The flux blanket shields the arc and the molten metal, preventing spattering and reducing atmospheric contamination.
- SAW is primarily used for welding thick materials and producing high-quality, consistent welds in various industries such as shipbuilding, heavy equipment, and pressure vessel fabrication.

These welding processes represent a range of methods available for different applications, materials, and welding conditions. By following AWS guidelines and techniques, welders can utilize the most suitable process and method to achieve strong, reliable, and defect-free welds in their respective industries.

Type of welds and weld joints

AWS guides different types of welds and weld joints used in various welding applications. Understanding these weld types and joints is crucial for ensuring proper joint design and achieving the desired weld quality. Here's an overview of some commonly recognized welds and weld joints as per AWS:

Types of Welds:

1. Fillet Weld: A fillet weld is a triangular weld joint formed between two pieces positioned perpendicular to each other. It is commonly used in corner joints, lap joints, and T-joints.

2. Groove Weld: A groove weld is created by filling a groove or channel between two workpieces. It may involve different joint configurations, such as single V, double V, U, J, or bevel groove.

3. Plug or Slot Weld: A plug or slot weld is formed by filling circular or elongated holes or slots in overlapping workpieces. They are commonly used for joining T-joints or lap joints.

4. Seam Weld: A seam weld is a continuous weld made along the length of two abutting workpieces. It is commonly used in joining sheets or plates in a lap joint configuration.

5. Spot Weld: A spot weld is formed by applying pressure and electric current to create a localized weld. It is typically used to join sheet metal or thin materials.

Types of Weld Joints:

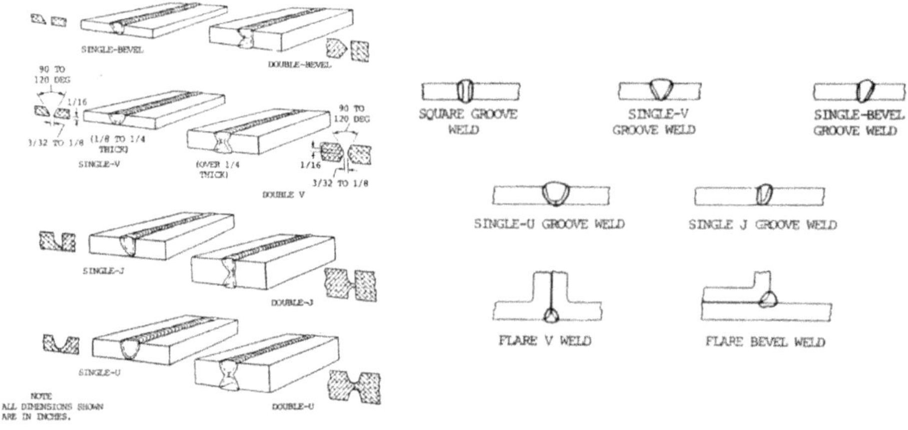

1. Butt Joint: In a butt joint, two workpieces are aligned edge-to-edge in the same plane to form a square or rectangular joint. Butt joints can be full penetration or partial penetration, depending on the weld depth.

2. Corner Joint: A corner joint occurs when two workpieces are joined at right angles to form a corner. Fillet welds are commonly used in corner joints.

3. Lap Joint: A lap joint is formed by overlapping two workpieces, typically with a fillet or plug weld. Lap joints provide increased strength and surface area for welding.

4. T-Joint: A T-joint forms when one workpiece is positioned at a right angle to another workpiece, resulting in a T-shaped configuration. Fillet welds are commonly used in T-joints.

5. Edge Joint: An edge joint joins two workpieces along their edges to form a seam, typically with a groove weld. Edge joints are commonly used in plate or sheet welding applications.

Understanding the different types of welds and weld joints allows welders to select appropriate joint designs, choose the right welding processes, and ensure the necessary strength and integrity of the welded components. Following AWS guidelines and standards helps ensure that welds and weld joints meet the required specifications and performance criteria.

Weld defects

As per the guidelines provided by AWS (American Welding Society), weld defects refer to imperfections or irregularities that may occur during the welding process, impacting the quality and integrity of the weld. Here's a brief overview of some commonly observed weld defects:

1. Porosity: Porosity appears as small gas pockets trapped within the weld or along the weld bead. It is caused by the presence of atmospheric gases, improper shielding gas coverage, or inadequate cleaning of the base metal.

2. Incomplete Fusion or Lack of Penetration: This defect occurs when the weld metal does not completely fuse with the base metal or fails to penetrate the joint fully. Inadequate heat input, improper joint preparation, or incorrect welding technique can contribute to this defect.

3. Lack of Fusion: Lack of fusion happens when there is inadequate bonding between the base metal and the weld metal. It can occur

either along the sidewalls of the joint or between multiple weld passes. Insufficient heat, improper welding technique, or improper joint fit-up can cause this defect.

4. Cracks: Cracks can manifest as visible separations within the weld or in the heat-affected zone. They can occur due to high levels of stress, rapid cooling rates, excessive heat input, or inadequate preheating.

5. Undercut: Undercut is a groove or depression formed along the edges of the weld. It can weaken the weld joint and increase the chances of failure. Undercutting can result from excessive welding current, improper electrode angle, or excessive travel speed.

6. Excessive Penetration: Excessive penetration occurs when the weld metal penetrates beyond the desired depth, often leading to burn-through or the formation of holes. It can be caused by excessive heat input, incorrect welding techniques, or improper selection of welding parameters.

7. Slag Inclusion: Slag inclusion refers to the entrapment of slag within the weld metal. It can occur when the weld bead does not completely fuse with the base metal, leaving voids or pockets that trap slag. Insufficient cleaning, improper electrode angle, or improper flux removal can cause this defect.

It is important to note that welding defects can affect the strength, structural integrity, and overall performance of the welded joint. Proper welding techniques, adherence to welding standards, and adequate quality control measures help minimize these defects and ensure high-quality welds. Regular inspection, testing, and quality assurance procedures are necessary to identify and rectify any weld defects, ensuring that the welded components meet the desired specifications and performance requirements.

Robotic welding

Robotic Welding: Revolutionizing Precision, Efficiency, and Quality

Robotic welding has emerged as a game-changer in the field of welding, revolutionizing the way metal components are joined. With its precision, efficiency, and consistent quality, robotic welding offers numerous advantages over traditional welding methods. Let's explore the world of robotic welding and its impact on various industries.

At the heart of robotic welding is the use of advanced robotic arms equipped with welding torches or manipulators. These robots are designed to perform complex welding tasks with unparalleled accuracy and speed. Guided by computer programming, they follow predefined paths to deposit the welded metal precisely, ensuring uniformity and minimizing human error.

One of the key advantages of robotic welding is its exceptional precision. The robot's repeatability and ability to maintain consistent parameters throughout the welding process result in highly accurate welds. This precision is especially critical in industries like automotive, aerospace, and manufacturing, where tight tolerances and exact specifications are required.

Efficiency is another significant benefit of robotic welding. Robots can work tirelessly, 24/7, allowing for increased productivity and reduced cycle times. They execute welds quickly and consistently, significantly improving production rates and reducing manufacturing lead times. With faster welding speeds, companies can meet market demands, streamline processes, and achieve higher output levels.

Robotic welding also offers unparalleled quality control. By incorporating sensors and cameras, robots can monitor and adjust parameters in real time, ensuring consistent weld quality. They detect variations such as gaps, distortion, or inconsistencies, making precise adjustments to maintain the integrity of the weld. This level of control minimizes defects and ensures a high-quality end product.

Safety is a key aspect of robotic welding. By automating the process, workers are less exposed to hazardous fumes, sparks, and high-temperature environments. The risk of accidents and injuries is significantly reduced, promoting a safer workplace environment. Human operators can focus on supervising and programming the robots, further enhancing safety protocols.

Not only does robotic welding result in higher efficiency and better quality, but it also enables cost savings. While the initial investment in robotic equipment is significant, the long-term benefits outweigh the upfront costs. Reduced labor expenses, increased productivity, and minimized rework contribute to a higher return on investment over time.

Robotic welding is versatile and applicable to various welding applications, including spot welding, arc welding, laser welding, and more. It can handle a wide range of materials, such as steel, aluminum, and stainless steel, catering to diverse industries like automotive, construction, fabrication, and beyond.

As technology advances, robotic welding continues to evolve, incorporating innovative features like machine learning and artificial intelligence. These advancements enable robots to adapt to changing welding conditions, detect anomalies, and optimize welding parameters in real time.

Robotic welding has revolutionized the welding industry, propelling it into a new era of precision, efficiency, and quality. In a highly competitive market, robotic welding plays a pivotal role in meeting customer demands, enhancing production capabilities, and maintaining consistent weld integrity. The possibilities are vast, and the future of robotic welding holds even more promise as technology continues to advance.

Welding procedure specification (WPS)

The Welding Procedure Specification (WPS) is a vital document in the fabrication industry that outlines the specific procedures and variables required to ensure proper welding operations. It serves as a guide for welders, inspectors, and other personnel involved in the fabrication process. Here's a brief overview of the WPS and its purpose:

1. Definition of the Welding Procedure Specification (WPS):

   — The WPS is a written document that details the welding procedures to be followed during fabrication. It defines the specific welding techniques, parameters, materials, and quality requirements necessary to achieve a successful weld.

2. Purpose of the Welding Procedure Specification:

   — Ensuring Consistency: The WPS provides a standardized approach to welding operations, ensuring consistency in welding procedures across various projects and fabrication practices.
   — Meeting Quality Standards: By following the guidelines outlined in the WPS, fabricators can meet the required quality standards and ensure that the welded joints meet or exceed the required mechanical and metallurgical properties.
   — Documentation and Compliance: The WPS serves as a documentation of the approved welding procedure, ensuring compliance with regulatory codes, industry standards, client specifications, and project requirements.
   — Welder Qualification: The WPS also serves as a basis for welder qualification. Welders are required to demonstrate their proficiency by welding test samples to the specifications outlined in the WPS to ensure their ability to produce sound welds.
   — Process Control and Safety: The WPS includes information regarding pre-weld/post-weld treatments, welding consumables, joint preparation, welding techniques, and any specific safety considerations. Adhering to the WPS helps maintain process control and ensures the safety of personnel involved in welding operations.

- Effective Communication: The WPS acts as a communication tool between different stakeholders involved in fabrication, such as designers, welding engineers, fabricators, and inspectors. It provides a clear understanding of the welding procedures and requirements to facilitate effective collaboration and minimize errors or misunderstandings.

In summary, the WPS plays a crucial role in achieving high-quality and consistent welds in fabrication. By following the procedures and variables outlined in the document, fabricators can ensure compliance with standards, meet project specifications, and produce welds that meet the required quality and structural integrity for the specific application.

Procedure qualification record (PQR)

The Procedure Qualification Record (PQR) is a crucial document in the fabrication industry that validates the soundness and reliability of a welding procedure. It serves as proof that a specific welding procedure, as outlined in the Welding Procedure Specification (WPS), successfully meets the required quality and performance criteria. Here's a brief overview of the PQR and its purpose:

1. Definition of the Procedure Qualification Record (PQR):

- The PQR is a documented record that contains data and test results from the qualification of a specific welding procedure. It demonstrates that the welding procedure, as defined in the WPS, produces welds of acceptable quality and mechanical properties.

2. Purpose of the Procedure Qualification Record:

- Quality Assurance: The PQR provides evidence that the welding procedure meets the required quality standards, ensuring that the resulting welds possess the desired mechanical and metallurgical properties.
- Compliance and Certification: The PQR serves as a means to demonstrate compliance with industry codes (e.g., AWS D1.1,

ASME IX) and specific project requirements. It is often a necessary prerequisite for obtaining certifications, permits, or contracts.

- Process Optimization: The PQR includes detailed information about the welding variables, such as heat input, filler metal, pre-heating, and interpass temperatures. By analyzing the results of the qualification tests, fabricators can optimize and refine the welding process, leading to improved efficiency and productivity.

- Welder Training and Qualification: The PQR is used for welder qualification, whereby welders must replicate the procedure outlined in the PQR. Successful qualification ensures that welders possess the necessary skills and proficiency to produce sound welds that meet the required standards.

- Documentation and Traceability: The PQR provides documented evidence of the welding procedure's successful qualification. It serves as an essential record for future reference, inspections, audits, and traceability purposes.

- Continuous Improvement: The PQR can also act as a reference point for improving existing welding procedures or developing new ones. By analyzing the PQR data, fabricators can identify areas for improvement, refine processes, and enhance the overall quality of the welds.

With its comprehensive documentation of testing results, the PQR plays a critical role in ensuring that welding procedures yield reliable, high-quality welds. By adhering to the requirements and specifications outlined in the PQR, fabricators can demonstrate compliance, improve weld quality, and instill confidence in the welding process for successful fabrication projects.

Welder qualification

Welder qualification is a crucial aspect in any fabrication shop to ensure that welders possess the necessary skills, knowledge, and proficiency to produce high-quality welds that meet the required standards. Here is an explanation of the welder qualification requirements typically applicable to a fabrication shop:

1. Certification Standards:

   – Identify the relevant certification standards that govern the fabrication industry, such as AWS D1.1 (structural welding) or ASME IX (boiler and pressure vessel welding).
   – Understand the specific requirements outlined in these standards, including welder qualification tests, documentation, and acceptance criteria.

2. Welder Performance Qualification (WPQ):

   – Determine the essential variables of the welding procedure to be used, such as the welding process, base materials, joint configurations, welding positions, and filler metals.
   – Prepare the Welder Performance Qualification (WPQ) test plate, ensuring it reflects the intended welding conditions and meets the standard's requirements.
   – Administer the WPQ test, which typically involves welding the test plate according to the specified procedure and passing visual inspections, non-destructive testing (NDT), and mechanical testing.

3. Test Documentation and Recordkeeping:

   – Document all relevant details of the welder qualification test, including welding procedure specifications (WPS), test plate identification, welding parameters, inspection results, and any additional testing performed.
   – Maintain accurate records of welder qualifications, ensuring traceability and compliance with industry standards.

4. Requalification and Retesting:

   – Establish requalification intervals and requirements for welders to ensure ongoing competency. This may include periodic retesting

or performance evaluations to validate continued skill and knowledge.
- Stay updated with changes in industry standards and specifications to ensure welders maintain the necessary qualifications.

5. Continual Improvement and Training:

- Encourage continual improvement through training programs, workshops, and certification courses to enhance welders' skills, knowledge, and understanding of new technologies and techniques.
- Promote a culture of safety and quality in the fabrication shop, emphasizing adherence to established procedures and best practices.

6. Inspections and Audits:

- Perform regular inspections and audits to verify that welders are following proper welding procedures and maintaining the required quality standards.
- Address any non-conformities or deficiencies promptly, taking corrective actions and providing additional training or support as needed.

By adhering to welder qualification requirements, fabrication shops can ensure that their welders possess the necessary skills and qualifications to execute welding operations confidently and produce welds that meet the required quality, safety, and industry standards. This commitment to welder qualification ultimately contributes to the overall success and reputation of the fabrication shop in delivering high-quality fabricated products.

All welding-related records are retained by the project quality control department.

Heat treatment (Stress relieving)

Different heat treatment processes are used after welding. also called post-weld heat treatment (PWHT)

Different Heat Treatment Processes for Stress Relieving After Welding

After welding, residual stresses can be present in the welded components, which can lead to distortion, cracking, or reduced mechanical properties. To alleviate these stresses and enhance the integrity of the welded structure, various heat treatment processes are employed for stress relieving. Here are some commonly used heat treatment processes after welding:

1. Post-Weld Heat Treatment (PWHT):

- PWHT involves heating the welded component to a specific temperature and maintaining it for a specified duration, followed by controlled cooling.
- The objective of PWHT is to reduce residual stresses, improve material properties, and enhance the weldment's ductility and toughness.
- The temperature and duration of PWHT are determined based on factors like material composition, welding procedure, and code or specification requirements.

2. Stress Relief Annealing:

- Stress relief annealing typically involves heating the welded component to a temperature below the lower critical transformation temperature and holding it for a specific time before slow cooling.
- This process aims to relieve thermal and residual stresses and restore the original mechanical properties of the material.
- Stress relief annealing is commonly applied to structures with thick sections or complex geometries where stresses may be more pronounced.

3. Vibratory Stress Relief (VSR):

- VSR is a non-thermal stress relieving technique that utilizes mechanical vibrations to reduce residual stresses.
- The welded component is subjected to controlled vibrations, which induce micro-plastic deformations, redistributing residual stresses and improving material stability.
- VSR is utilized to relieve stresses in sensitive or heat-sensitive materials or applications where thermal treatment may not be suitable.

4. Induction Heating:

- Induction heating involves using electromagnetic fields to induce eddy currents in the welded component, resulting in localized heating.
- By selectively heating specific areas, the localized thermal expansion helps to relieve residual stresses.
- Induction heating is particularly suitable for relieving stresses in thick sections or near critical areas that are susceptible to cracking.

5. Stress Relief with Cryogenic Treatment:

- Cryogenic treatment involves subjecting the welded component to extremely low-temperature environments, typically below -150°C (-238°F).
- Slow cooling and maintaining the temperature at cryogenic levels allow for stress reduction and stabilization of the material structure.
- Cryogenic treatment is especially useful in enhancing the mechanical properties and dimensional stability of certain materials, such as tool steels.

The selection of the appropriate heat treatment process for stress relieving after welding depends on factors such as material type, weldment design, welding procedure, and desired results. Consulting relevant codes, standards, and specifications can provide guidelines for determining the specific heat treatment requirements. Properly executed stress-relieving processes contribute to the longevity, reliability, and performance of welded structures by minimizing residual stresses and potential failure mechanisms.

Non-Destructive examination

Non-Destructive Examination (NDE) Techniques: Ensuring Integrity and Safety

Non-Destructive Examination (NDE) techniques play a critical role in verifying the integrity and safety of equipment, structures, and materials without causing damage or impairing their functionality. These methods are widely applied in industries such as oil and gas, aerospace, manufacturing, and infrastructure. This write-up explores various NDE techniques commonly used to assess quality, detect defects, and ensure the reliability of critical components.

1. Ultrasonic Testing (UT):

Ultrasonic Testing utilizes high-frequency sound waves to assess the internal structure of materials. It involves transmitting ultrasonic pulses into a component and analyzing the reflected waves to identify defects such as cracks, voids, and weld discontinuities. UT is particularly effective in determining thickness measurements, flaw detection, and evaluating the integrity of materials such as metals, composites, and concrete.

2. Radiographic Testing (RT):

Radiographic Testing employs X-rays, gamma rays, or other ionizing radiation to inspect materials for hidden defects or abnormalities. The process involves exposing a material to radiation and capturing the resulting image on a film or digital detector. RT is commonly used to evaluate welds, castings, forgings, and other metallic components to detect internal flaws, porosity, or inclusions that may compromise their integrity.

3. Magnetic Particle Testing (MT):

Magnetic Particle Testing utilizes magnetic fields and finely divided ferromagnetic particles to detect surface and near-surface defects in ferromagnetic materials. The technique involves magnetizing the component while applying magnetic particles. Any defects or cracks present in the material disrupt the magnetic field, causing the particles to concentrate at the defect, thus making it visible under proper lighting conditions.

4. Liquid Penetrant Testing (PT):

Liquid Penetrant Testing is utilized to identify surface-breaking defects, such as cracks and porosity, in non-porous materials. The process involves applying a liquid penetrant to the surface of the component, allowing it to seep into any surface openings or defects. The excess penetrant is then removed, and a developer is applied to draw out the penetrant, making the defects visible for inspection.

5. Eddy Current Testing (ET):

Eddy Current Testing uses electromagnetic induction to assess the electrical and mechanical properties of conductive materials. The technique involves passing an alternating current through a coil, creating an oscillating magnetic field. Any variations within the material, such as changes in conductivity or defects, cause a change in the eddy currents induced in the material. This change is detected and analyzed to identify flaws or material characteristics.

6. Visual Inspection (VI):

Visual Inspection is a fundamental technique used to assess the external condition of components and structures. It involves visually examining surfaces, welds, joints, and connections for abnormalities, discontinuities, or signs of degradation. VI is often the first line of defense in quality control, providing a quick but essential assessment of the overall condition or cleanliness of a component

Destructive testing

Destructive testing plays a crucial role in the quality assurance processes of a fabrication shop. This form of testing involves subjecting materials, components, or weldments to rigorous tests that ultimately result in their failure. While the term "destructive" may sound negative, these tests are essential for validating structural integrity, identifying weaknesses, and ensuring the overall safety and reliability of fabricated products. Here is a breakdown of some common destructive testing methods performed at a fabrication shop:

1. Tensile Testing:

- Tensile testing evaluates the mechanical properties of materials by subjecting them to axial pulling forces until they fracture. This test measures parameters such as ultimate tensile strength, yield strength, elongation, and modulus of elasticity. It helps verify material conformity, strength, and suitability for specific applications.

2. Bend Testing:

- Bend testing evaluates the ductility and soundness of welded or fabricated joints. The test involves bending a standardized specimen to a specified angle without causing fractures or defects. It helps determine weld quality, such as the absence of cracks, lack of fusion, or other discontinuities that could compromise structural integrity.

3. Charpy Impact Testing:

- Charpy impact testing assesses the ability of materials to resist brittle fracture under impact loading. A notched test specimen is struck by a pendulum, and the energy absorbed during fracture is measured. The results indicate the material's toughness and ability to withstand sudden impacts, crucial for assessing material suitability in structural applications

4. Hardness Testing:

- Hardness testing measures the material's resistance to penetration or deformation. Common methods include the Rockwell, Brinell, and Vickers hardness tests. These tests evaluate material strength, wear resistance, and suitability for specific operational requirements.

5. Fatigue Testing:

- Fatigue testing subjects materials or components to cyclic loading to evaluate their durability under repetitive stress. This test helps determine a material's resistance to cracks, fractures, and failure caused by repeated loading and unloading cycles, simulating real-world operational conditions.

6. Weld Testing:

- Various destructive tests are performed on welded joints to assess their quality and conformity to industry standards.

Examples include tensile testing, bend testing, macro- and micro-examinations, visual inspections, and radiographic testing. These tests help ensure the integrity of welds, identifying potential defects, discontinuities, or weaknesses that could compromise structural integrity.

It's important to remember that destructive testing is only performed on representative samples, rather than the entire fabricated product. The goal is to obtain critical information about materials, weldments, or components to ensure they meet specified requirements and comply with applicable codes, standards, and client expectations.

By conducting these destructive tests, fabrication shops can maintain quality control, demonstrate compliance, and provide clients with products that are safe, reliable, and built to withstand the intended operational demands.

Project Control

Project Controls, Planning, and Cost Control are essential elements within the EPC (Engineering, Procurement, and Construction) sector of the oil and gas industry. These functions ensure effective project management, efficient resource allocation, and cost optimization throughout the project lifecycle.

1. Project Controls: Project Controls encompass a set of processes and tools used to monitor, control, and report on various aspects of a project. It includes the integration of project planning, scheduling, cost estimation, budgeting, risk management, and performance measurement. Project Controls aim to provide accurate and timely information to stakeholders, enabling them to make informed decisions and ensure successful project delivery.

2. Planning: Planning is a critical component of project management that involves defining project objectives, identifying activities,

establishing project milestones, and creating timelines. In the EPC industry, planning involves developing a comprehensive project execution plan outlining the sequence of activities, resource requirements, and coordination among different engineering disciplines, procurement, construction, and Commissioning & Start-up (CSU) phases.

3. Cost Control: Cost Control is the process of monitoring and managing project expenditures to align with the approved budget. In the oil and gas EPC industry, cost control involves tracking costs related to engineering design, procurement of materials and equipment, construction activities, labor, subcontracting, and other project-related expenses. Effective cost control measures aim to avoid budget overruns, optimize resource utilization, and achieve cost savings without compromising project quality and schedule.

These functions work together to ensure project success:

- Project Controls provide oversight and coordination of project activities, integrating planning, scheduling, and cost control to maintain project alignment with objectives and stakeholder expectations.
- Planning establishes a roadmap for project execution, enabling efficient resource management, sequencing of activities, and coordination of multiple stakeholders.
- Cost Control ensures effective cost estimation, expenditure tracking, and cost analysis, supporting the development and maintenance of a realistic project budget.

By implementing robust project controls, effective planning, and efficient cost control measures, EPC projects in the oil and gas industry can minimize risks, maximize operational efficiency, and deliver projects on time and within budget. These practices help in meeting project goals, quality standards, and client expectations while adhering to industry regulations and safety guidelines.

Project Document Control department

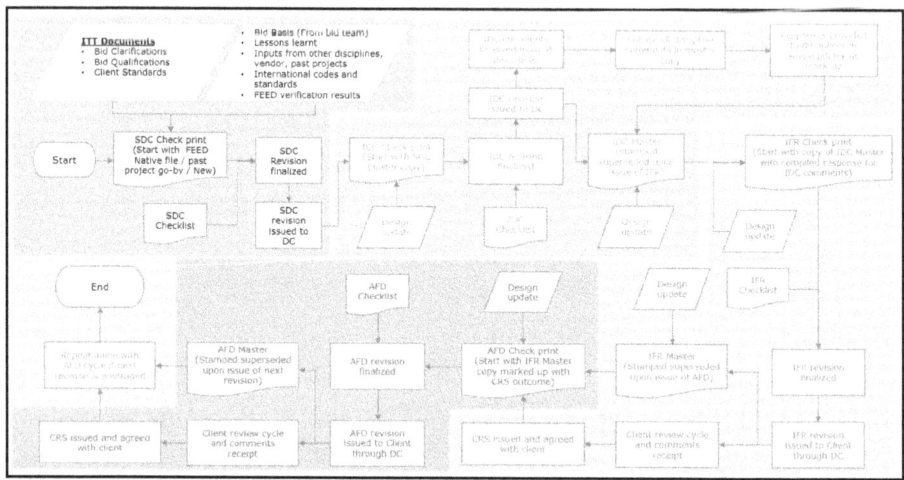

Project document control is a critical aspect of EPC (Engineering, Procurement, and Construction) projects that ensure the proper management, organization, and control of project-related documents and information throughout the project lifecycle. It involves establishing efficient processes, systems, and protocols for document creation, distribution, tracking, storage, and retrieval.

In the realm of EPC projects, the importance of effective project document control can be summarized as follows:

1. Document Management: Project document control establishes a systematic approach to managing documents, including drawings, specifications, contracts, reports, correspondence, and other project-related records. It ensures that the right documents are available to the right stakeholders at the right time, facilitating smooth project execution and decision-making.

2. Version Control: Document control ensures that project documents are up-to-date, accurate, and consistent. It manages document revisions, ensuring that the latest versions are accessible and that obsolete documents are properly archived or disposed

of. This minimizes confusion, mitigates risks associated with using outdated information, and facilitates efficient collaboration among project teams.

3. Information Flow and Distribution: Document control governs the distribution of project documents, ensuring that information is shared appropriately with authorized stakeholders. It establishes protocols for document transmittals, review cycles, approvals, and notifications, facilitating efficient communication and timely access to critical information.

A document Distribution Matrix is prepared for each project for distributing the documents to the relevant stakeholders, for either their review or for information.

There are many EDMS (Electronic document management system) software available in the market to manage the workflow.

4. Change Management: Effective document control is essential for managing changes within the project. It ensures that change requests, revisions, and modifications to documents are properly documented, tracked, reviewed, and approved. This helps maintain project scope, minimize potential conflicts, and ensure compliance with regulatory requirements or client specifications.

5. Compliance and Audit Readiness: Document control is vital for maintaining compliance with industry standards, regulations, contractual obligations, and quality management systems. It enables the tracking of project-related documentation and facilitates audits, inspections, and regulatory reviews. Proper document control documentation also aids in demonstrating adherence to project requirements and documentation standards.

6. Knowledge Management: Through document control, project documents become part of the organization's knowledge base. They serve as valuable references for future projects, post-project

analysis, continuous improvement, and lessons learned. Properly organized and managed project documentation streamlines knowledge transfer, allowing organizations to build upon past experiences and improve project outcomes.

Efficient project document control contributes to enhanced communication, collaboration, and coordination within EPC projects. It ensures that stakeholders have access to accurate and timely information, facilitates compliance with regulatory and contractual obligations, minimizes risks associated with outdated or incorrect data, and supports effective decision-making throughout the project lifecycle.

Critical path

The critical path in EPC (Engineering, Procurement, and Construction) project planning refers to the sequence of activities that determines the project's overall duration. It is the longest path from the start to the end of the project, which means that any delay in activities on the critical path will directly impact the project timeline.

The critical path is identified by analyzing the project's network diagram or using project scheduling software. It consists of activities that have zero float or slack, meaning they have no flexibility in their start or end dates. If any of the activities on the critical path are delayed, the project's completion date will be affected.

By focusing on the critical path, project managers can allocate resources and prioritize activities to ensure timely project completion. They closely monitor the progress of critical path activities, identify potential bottlenecks, and take necessary actions to mitigate risks and keep the project on schedule.

Cost Management

Cost management during project execution involves the processes and activities required to estimate, budget, track, control, and manage project costs effectively throughout the project's lifecycle. Here is an overview of how cost management typically occurs during project execution:

1. Cost Estimation:

   - At the beginning of the project, cost estimation is performed to determine an approximate budget for the project. This estimation is based on the project scope, available information, historical data, and expert judgment. Various estimation techniques, such as analogous estimating or parametric estimating, may be used to assess project costs.

2. Cost Budgeting:

   - Once the cost estimation is complete, the project manager and stakeholders develop a comprehensive budget plan. This plan allocates the estimated costs to different project activities, work packages, or work breakdown structure (WBS) elements. The budget includes labor costs, material costs, equipment costs, contingency reserves, and any other relevant expenses.

3. Cost Tracking:

   - During the project execution phase, project managers closely monitor and track actual costs incurred against the budgeted costs. This involves regularly recording and analyzing expenditure data, invoices, labor hours, and other financial metrics. Tracking can be done manually or through project management software, and it helps identify any deviations from the budget.

4. Cost Control:

   - Cost control is a crucial aspect of project execution. It involves taking proactive measures to manage costs within the approved budget. If there are unforeseen cost variances or budget overruns, project managers analyze the causes of the variances, assess their impact on project objectives, and implement corrective actions. Cost control measures can include renegotiating contracts, adjusting scope, reallocating resources, or revising the project plan.

5. Change Management:

- Project changes, such as scope changes, revised requirements, or adjustments to the project plan, can affect costs. During project execution, it is important to assess the impact of changes on the budget and overall project costs. Change management processes are implemented to evaluate change requests, estimate their effects on costs, and obtain appropriate approvals before implementation.

6. Earned Value Management (EVM):

- Earned Value Management is a technique used to assess project performance, progress, and cost efficiency. It integrates schedule, cost, and performance metrics to provide insight into the project's earned value, actual costs, and planned costs. By analyzing key EVM indicators like schedule variance (SV), cost variance (CV), and cost performance index (CPI), project managers can identify cost-related issues and take corrective actions if necessary.

7. Reporting and Communication:

- Throughout project execution, project managers provide regular cost status reports to stakeholders. These reports summarize the project's financial status, including budgeted costs, actual costs, variances, and forecasts. Effective communication ensures that stakeholders are updated on cost-related matters and that any necessary decisions or actions can be taken.

By following these cost management practices during project execution, organizations can maintain control over project expenditures, ensure alignment with budgets, and monitor the financial health of the project in real time. This allows for efficient resource allocation, timely decision-making, and successful delivery of the project within the approved cost parameters.

Interface management

Interface management in EPC project execution in the oil and gas sector refers to the systematic process of identifying, defining, coordinating, and controlling the interfaces between various stakeholders, disciplines, systems, and activities within the project. It aims to ensure seamless integration, effective communication, and collaboration between multiple parties involved in the project to minimize conflicts, prevent errors, and maximize project performance.

In the context of an EPC project in the oil and gas sector, interface management focuses on the coordination and integration of the following key aspects:

1. Stakeholders: Interface management involves identifying and managing the interfaces between different project stakeholders, including the project owner, EPC contractor, subcontractors, suppliers, regulatory bodies, local communities, and other relevant parties. It ensures effective communication, aligns expectations, and resolves any conflicts or issues that may arise during the project execution.

2. Engineering Disciplines: An EPC project typically involves a multitude of engineering disciplines such as civil, mechanical, electrical, instrumentation, and process engineering. Interface management ensures proper coordination and collaboration between these disciplines to ensure that design requirements, specifications, and interfaces are properly defined and integrated. It helps prevent design clashes and optimizes the overall engineering process.

3. Systems and Processes: In an EPC project, various systems and processes are interconnected and dependent on one another. Interface management ensures effective coordination and integration of these systems and processes, such as procurement, construction, commissioning, and operation. It ensures that handovers, interfaces, and dependencies are properly defined, managed, and monitored throughout the project lifecycle.

4. Interfaces between Subsystems: Large-scale EPC projects in the oil and gas sector often consist of several subsystems that need to be seamlessly integrated. Interface management focuses on identifying, defining, and coordinating the interfaces between these subsystems (e.g., process systems, mechanical systems, electrical systems) to ensure smooth integration and functionality.

5. Communication and Documentation: Effective communication and documentation are essential for successful interface management. It involves establishing communication channels, protocols, and information exchange mechanisms to ensure that relevant project information is shared accurately and promptly. It also includes managing and maintaining interface documentation, such as interface control documents, interface registers, and interface requirements.

6. Risk Management: Interface management plays a crucial role in identifying and managing risks associated with interface dependencies. It includes proactive identification of potential interface conflicts or issues, defining risk mitigation measures, and monitoring the implementation of these measures to minimize the occurrence and impact of interface-related risks.

By implementing effective interface management practices, EPC projects in the oil and gas sector can enhance collaboration, streamline processes, minimize conflicts, and improve overall project performance. It helps to ensure that different project elements are properly integrated, align with project objectives, and contribute to the successful execution of the project.

1. Internal Interface Management:

Internal interface management focuses on coordinating and integrating various elements within the project organization itself. This may include the following aspects:

- Interdisciplinary Coordination: Ensuring effective coordination among different engineering disciplines within the project, such as civil, mechanical, electrical, instrumentation, and process engineering. This involves defining interface requirements, coordinating design activities, and resolving any conflicts or clashes between different disciplines.
- Project Stakeholder Coordination: Managing interfaces and interactions between different project stakeholders within the organization, such as project management, engineering team, procurement team, construction team, and subcontractors. This ensures effective communication, collaboration, and coordination among internal stakeholders to achieve project objectives.
- System Integration: Coordinating the integration of various systems and processes within the project, such as procurement, construction, commissioning, and operation. This involves managing handover points, dependencies, and interfaces between different project stages, and ensuring seamless integration of these systems to achieve project milestones.
- Documentation and Communication: Establishing effective communication channels and managing documentation to facilitate the exchange of information and ensure that all internal stakeholders have access to relevant project information. This includes maintaining interface control documents, interface registers, and other documentation to track and manage internal interfaces.

2. External Interface Management:

External interface management focuses on coordinating and managing interfaces with external stakeholders and systems beyond the project organization. This may include the following aspects:

- Client and Contractor Coordination: Coordinating interfaces between the project owner (client) and the EPC contractor, including managing project requirements, expectations, and contractual obligations. It also involves aligning project objectives, resolving conflicts, and facilitating effective communication between the client, contractor, and other external stakeholders.
- Subcontractor Management: Managing interfaces with subcontractors, including defining requirements, coordinating activities, and monitoring their performance to ensure alignment with the overall project schedule and objectives. This involves effective communication, regular progress monitoring, and timely issue resolution between the main contractor and subcontractors.
- Regulatory Compliance: Coordinating interfaces with regulatory bodies, government agencies, and local communities to ensure compliance with applicable laws, regulations, and permitting requirements. This includes obtaining necessary approvals, permits, and clearances, as well as addressing any interface-related issues or concerns raised by external stakeholders.
- Supply Chain Coordination: Managing interfaces with suppliers, vendors, and logistics providers to ensure timely delivery of materials, equipment, and services. This involves coordinating procurement activities, tracking orders, managing transportation logistics, and maintaining effective communication with external suppliers to minimize delays and disruptions.
- Interface with Other Projects: In certain cases, EPC projects may be part of a larger program or involve interfaces with adjacent projects. Managing these interfaces requires coordination and

collaboration to ensure seamless integration, alignment of project activities, and resolution of any inter-project conflicts or issues.

Effective internal and external interface management is crucial for the successful execution of EPC projects. It ensures smooth coordination, effective communication, and seamless integration among various stakeholders, systems, and processes both within and outside the project organization.

In a typical EPC project, several interfaces between Engineering, Procurement, Construction, and Commissioning (EPCC) need to be effectively managed. Here are the key interfaces between these four phases:

1. Engineering and Procurement:

- Engineering Specifications and Procurement: The engineering phase defines the technical specifications and requirements for materials and equipment. The procurement phase relies on these specifications to identify suitable suppliers and procure the required materials and equipment.
- Engineering and Supplier Technical Clarifications: During procurement, suppliers may seek clarifications or technical inputs from the engineering team to ensure their offerings meet the project requirements. Close collaboration and communication between engineering and procurement teams are necessary to address any technical queries or issues.
- Engineering Design and Procurement Delivery Schedule: The engineering design phase establishes the project schedule, including milestones for the delivery of detailed engineering drawings and data to support procurement. The procurement team needs these deliverables to procure materials and equipment within the desired timeframes.

2. Procurement and Construction:

- Procurement Schedule and Construction Planning: The procurement schedule, including material delivery timelines, plays a crucial role in construction planning. The construction team needs to know when materials and equipment will be available on-site to schedule manpower, logistics, and construction activities accordingly.

- Material Logistics and Construction Site Access: The procurement team must coordinate with the construction team to facilitate the seamless delivery of materials and equipment to the construction site. This includes ensuring smooth logistics, providing necessary documentation for transportation, and coordinating site access for material offloading.

- Quality Assurance and Inspections: The procurement team collaborates with the construction team to conduct quality inspections and ensure adherence to project specifications. This requires clear communication and coordination to address any quality issues or non-conformities during construction.

3. Construction and Commissioning:

- Construction Handover and Commissioning Preparation: As construction nears completion, there must be a smooth handover from the construction team to the commissioning team. This involves transferring crucial documentation, identifying pending actions, and preparing the construction site for commissioning activities.

- Documentation and System Acceptance: The construction team provides critical documents and as-built drawings to the commissioning team for system acceptance and handover. Clear communication and proper documentation exchange are vital to ensure a smooth transition and avoid delays or confusion during the commissioning phase.

– Construction Punch-list Resolution: In the commissioning phase, the construction team works with the commissioning team to address any outstanding construction punch-list items before commissioning activities can proceed. Collaboration and coordination are necessary to rectify any remaining construction issues and ensure a robust start to the commissioning process.

Effective management of these interfaces between Engineering, Procurement, Construction, and Commissioning is essential for the seamless execution of an EPC project. It requires clear communication, coordination, and collaboration among these functional teams to ensure a smooth flow from one phase to another and ultimately achieve project success.

Contract management

Contract management plays a crucial role in the EPC (Engineering, Procurement, and Construction) sector of the oil and gas industry. It involves the effective administration and management of contractual agreements between project owners, contractors, and suppliers. Contract management is essential for ensuring compliance with legal requirements, mitigating risks, and achieving successful project outcomes. Here is a brief overview of Contracts Management in the EPC oil and gas industry:

1. Contract Formation: Contracts in the EPC industry are typically complex, addressing various aspects such as scope of work, specifications, deliverables, milestones, pricing, and legal terms. Contracts Management oversees the process of contract formation, involving the negotiation, drafting, review, and finalization of contractual agreements. It ensures that contracts accurately capture the expectations and obligations of all parties involved.

2. Risk Management: Contracts Management integrates risk identification, assessment, and mitigation strategies into

contracts. This involves defining outcome-based performance indicators, allocating risks and liabilities, and establishing dispute resolution mechanisms. Effective risk management in contracts helps protect the interests of all parties and ensures smooth project execution.

3. Contract Administration: Once contracts are in place, Contracts Management focuses on contract administration, which includes monitoring compliance with contractual obligations, tracking project progress, and managing change orders. It involves facilitating communication, resolving issues, and ensuring proper documentation of contract-related activities.

4. Contractual Variations: In the EPC industry, changes to project scope or specifications may arise during a project. Contracts Management oversees the management of contractual variations, which involves reviewing change requests, negotiating costs and timelines, and amending contracts appropriately. Effective management of variations helps maintain project momentum and minimize delays or disputes.

5. Claims and Dispute Resolution: In cases where conflicts or claims arise, Contracts Management plays a vital role in managing and resolving disputes. This involves conducting investigations, facilitating negotiation and mediation processes, and potentially engaging in formal dispute resolution mechanisms such as arbitration or litigation. Contract management strives to find equitable solutions that protect the interests of all parties involved.

6. Contract Closeout: Upon completion of the project, Contracts Management oversees the contract closeout process. This involves verifying the completion of deliverables, conducting final inspections, settling outstanding payments or claims, and ensuring proper documentation for audits or future reference.

Effective contract management in the EPC oil and gas industry ensures that contractual obligations are met, risks are mitigated, and projects are executed successfully. It requires in-depth knowledge of contract law, project management principles, and industry-specific regulations. By emphasizing transparency, collaboration, and effective communication, contract management contributes to the overall efficiency, profitability, and reputation of EPC projects in the oil and gas sector.

Project management consultant (PMC)

Project management consultant as an organization rather than an individual, the responsibilities remain similar, albeit at a broader level. As an organization providing project management consulting services in a greenfield refinery project, your role encompasses the following:

1. Project Planning and Strategy: As a project management consultancy, you will work closely with the client to develop a comprehensive project plan and strategy aligned with their objectives. This includes defining project scope, setting goals, creating timelines, and determining resource requirements. Your organization will provide strategic guidance to ensure the project's successful execution.

2. Stakeholder Engagement and Communication: Your organization will take the lead in engaging and managing stakeholders throughout the project. This involves establishing effective communication channels, facilitating collaboration among stakeholders, and addressing any concerns or conflicts. You will act as a central point of contact and ensure that all stakeholders are well-informed and aligned with project goals.

3. Risk Management and Mitigation: Your organization will conduct thorough risk assessments to identify potential risks and develop strategies to mitigate them. This includes analyzing project-related risks, developing risk management plans, and implementing risk mitigation measures. You will collaborate with the project team to monitor and manage risks throughout the project lifecycle.

4. Project Governance and Oversight: As a project management consultancy, you will provide overall governance and oversight for the project. This includes establishing project governance structures and frameworks, defining roles and responsibilities, and ensuring adherence to project management standards and best practices. Your organization will monitor project progress, review performance, and guide to ensure project success.

5. Quality Assurance and Control: Your organization will establish rigorous quality assurance processes to ensure that project deliverables meet the required standards. This includes setting up quality control mechanisms, conducting inspections and audits, and implementing corrective actions as needed. You will work closely with the project team to uphold quality standards throughout the construction of the greenfield refinery.

6. Cost and Budget Management: Your organization will be responsible for managing project finances, including cost estimating, budgeting, and cost tracking. You will collaborate with the client and project team to develop and monitor the project budget, analyze cost variances, and implement cost control measures. Your organization will provide financial reporting and forecasting to support effective resource allocation and cost efficiency.

7. Change and Risk Management: Your organization will take a proactive approach to change management, assessing the impact of change requests, and ensuring proper documentation and communication. Additionally, you will continuously monitor and manage project risks, identifying potential risks and implementing risk mitigation strategies.

8. Knowledge Management and Documentation: Your organization will establish processes for knowledge management and documentation to ensure comprehensive project records. This includes maintaining project documentation, creating progress

reports, and capturing lessons learned. Your organization will ensure that valuable project knowledge is captured and can be utilized for future projects.

As an organization providing project management consulting services in a greenfield refinery project, your role encompasses strategic planning, stakeholder management, risk mitigation, project governance, quality assurance, cost management, change management, knowledge management, and documentation. Your expertise and guidance will contribute to the overall success and timely completion of the project.

Project Risk Management

Project risk management is the art and science of managing risks caused by unforeseen changes (uncertainties) which may require deviations from the planned approach and may therefore affect the achievement of the project objectives. It involves systematically identifying, analyzing, planning, and controlling risks.

What is Risk?

A situation involving exposure to danger, loss, or harm.

Risk analysis
Identify risk
Analysis of risk

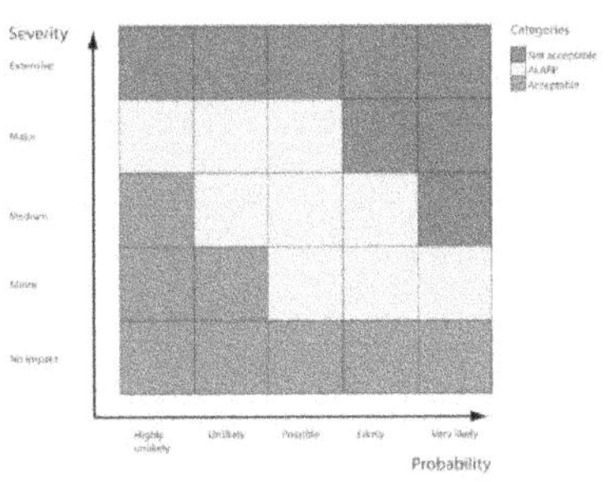

An unforeseen situation leading to an event, the occurrence of which is likely to deviate from the estimated or forecast value or planned path.

The probability of occurrence of the event.
Likely impact of that event, i.e., loss or gain
Mitigation of risk

If you don't consider Risk…. Remember you are at Risk….

External Environments Predictable Sources	Internal Environment		External Environments Unknown uncertainties
Political and Legal	Scope change, Time overrun, Cost Overrun	Leadership and organizational failure	Acts of God
Design and Specification	Technology Change	Resource Failure	Ecology
Financial and Economical	Quality and specification failure	Contractor Failure	Safety and Health

S. no	Risk Description	S.no	Risk Description
1	Project Scope Risk / Feasibility Risk	11	Force majeure & ecological Risk
2	Design & Specification Risk	12	Political, Legal and Social Risks
3	Quality Risk	13	Financial & Economical Risk
4	Time overrun Risk	14	Safety Health and environmental Risk
5	Cost Overrun Risk	15	Funding Failure Risk
6	Leadership Risk	16	Communication and network failure Risks
7	Organizational Risk	17	Project Execution Risks
8	Physical Resource mobilization and utilization Risk	18	Installation Risks of Mechanical and Electrical works.
9	Technology Risk	19	Purchase and Procurement Risks
10	Contractual Risk		

Project risk management in the EPC (Engineering, Procurement, and Construction) sector of the oil and gas industry involves identifying, assessing, mitigating, and monitoring risks that may impact project objectives, timelines, costs, and quality. It is a proactive and systematic approach that aims to ensure effective risk mitigation and project success.

1. Risk Identification: The first step in project risk management is identifying potential risks that may arise during the project lifecycle. This includes conducting a comprehensive assessment of various factors such as project scope, technical challenges, market conditions, political and regulatory risks, environmental factors, safety considerations, and stakeholder dynamics.

2. Risk Assessment and Prioritization: Identified risks are then assessed to determine their potential impact and likelihood of occurrence. The risks are prioritized based on their significance, allowing project stakeholders to focus on addressing high-priority risks that may have a significant impact on project objectives.

3. Risk Mitigation and Response Planning: Risk mitigation strategies are developed to address identified risks. Mitigation efforts involve creating contingency plans, implementing

relevant controls, and assigning responsibilities to ensure effective risk management. Response plans are developed to outline specific actions and procedures to be followed in the event a risk materializes.

4. Monitoring and Control: Throughout the project lifecycle, project risk management involves continuous monitoring and control of identified risks. This includes tracking risk status, reassessing the impact and likelihood of risks as the project progresses, and taking necessary actions to control and mitigate risks promptly. Regular risk reviews, progress assessments, and risk tracking tools are used to facilitate effective risk monitoring and control.

5. Lessons Learned and Continuous Improvement: Project risk management includes capturing and applying lessons learned from past projects to improve risk management practices. Risk management processes are periodically reviewed, and feedback is integrated to enhance risk identification, assessment, and mitigation strategies for future projects. This promotes a culture of continuous improvement in risk management within the organization.

6. Stakeholder Engagement: Effective project risk management involves engaging stakeholders throughout the process. This includes fostering open communication channels, sharing risk information and updates, and seeking input and feedback from stakeholders. Engaging stakeholders helps identify additional risks, incorporate diverse perspectives, and collectively develop risk mitigation strategies.

Project risk management in the EPC oil and gas sector is crucial for successful project execution. By identifying and mitigating risks through proactive planning, monitoring, and control, organizations can enhance project outcomes, optimize resource allocation, minimize delays, and achieve project objectives while adhering to industry standards and regulatory requirements.

Project change management

Project change management in the EPC (Engineering, Procurement, and Construction) sector of the oil and gas industry involves effectively managing and controlling changes that may arise during the project lifecycle. It encompasses processes and procedures to handle modifications, deviations, or variations to project scope, schedule, cost, specifications, or other project parameters.

1. Change Identification: Change management begins with the identification of potential changes that may impact the project. This includes changes in project requirements, design modifications, client requests, unforeseen issues, regulatory updates, or external factors that may arise during project execution.

2. Impact Assessment: Identified changes are then assessed for their potential impact on project scope, schedule, cost, quality, risks, and other parameters. The assessment considers the implications of each change on project objectives, stakeholders, resources, and deliverables.

3. Change Request and Evaluation: Based on the impact assessment, change requests are formally made, documented, and evaluated. Change requests should include a clear description of the proposed change, rationale, potential benefits or drawbacks, and the resources required for implementation.

4. Change Approval and Authorization: Change requests are reviewed and assessed by project stakeholders, including the project manager, client, and other relevant parties. Approval and authorization processes are followed to determine whether the change should be approved, rejected, or deferred for further evaluation.

5. Change Implementation and Communication: Once a change is approved, it is incorporated into the project plan. The change is communicated to relevant team members, subcontractors,

and stakeholders. Necessary adjustments are made to project schedules, budgets, resources, contractual agreements, and other project elements affected by the approved change.

6. Change Tracking and Documentation: Changes are systematically tracked and documented throughout the project lifecycle. This includes maintaining a change register, changelog, or change management system to record details of the change, its impact, approval status, implementation date, and any associated actions or decisions.

7. Change Control and Monitoring: Change control procedures are established to monitor and control changes effectively. This involves ongoing tracking of approved changes, assessing their impact on project progress and performance, managing related risks, and ensuring adherence to change management processes.

8. Lessons Learned and Continuous Improvement: Project change management incorporates capturing and applying lessons learned from change management activities. Feedback from past projects and change management experiences is utilized to enhance change management processes, controls, and stakeholder engagement for future projects.

Efficient project change management in the EPC oil and gas sector helps organizations respond to evolving project requirements, mitigate risks, maintain control, and ensure project success. By following structured change management processes, organizations can adapt to changes effectively while minimizing disruption, delays, cost overruns, and potential conflicts.

Crisis management

Crisis management in the EPC (Engineering, Procurement, and Construction) sector of the oil and gas industry involves effectively responding to and mitigating unexpected events, emergencies, or situations that pose a significant threat to the project or the safety and well-being of stakeholders. It encompasses a set of proactive measures and practices aimed at minimizing the impact of crises and ensuring the resilience and recovery of project operations.

1. Risk Identification and Preparedness: Crisis management begins with identifying potential risks and conducting risk assessments to anticipate potential crises. This includes identifying various scenarios, analyzing their potential impacts, and developing contingency plans and response strategies.

2. Emergency Response Planning: A key aspect of crisis management is the development of comprehensive emergency response plans. These plans outline clear procedures, roles and responsibilities, communication channels, and protocols for responding to different types of emergencies such as natural disasters, accidents, safety breaches, or security threats.

3. Incident Monitoring and Communication: During a crisis, monitoring the situation in real time is crucial. Crisis management involves establishing effective communication channels and systems to relay incident information promptly to relevant stakeholders, including project teams, clients, contractors, government authorities, and community representatives.

4. Resource Coordination and Mobilization: Crisis management requires the efficient coordination and mobilization of resources to address the crisis effectively. This includes activating emergency response teams, arranging necessary equipment, engaging relevant contractors, and ensuring the availability of necessary supplies, services, and expertise to manage the crisis.

5. Stakeholder Engagement and Communication: Crisis management involves engaging and communicating with stakeholders throughout the crisis event. Regular updates, clear and concise communication, establishing a central point of information, and addressing concerns and questions are vital components of effective crisis communication.

6. Lessons Learned and Continuous Improvement: Crisis management incorporates capturing lessons learned from past crises to strengthen future crisis response. Evaluating the effectiveness of crisis response measures, conducting post-crisis reviews, and implementing improvements in emergency response plans and strategies contribute to continuous improvement in crisis management practices.

7. Regulatory Compliance and Reporting: Crisis management in the oil and gas sector includes adherence to regulatory requirements and reporting obligations. This involves complying with safety, environmental, and security regulations, and reporting incidents or emergencies promptly to regulators and authorities as required.

Efficient crisis management in the EPC oil and gas sector emphasizes preparedness, rapid response, clear communication, and effective resource management. By implementing robust crisis management strategies, organizations can minimize the impact of crises, protect lives and assets, and quickly recover and resume normal project operations.

Business continuity plan

A business continuity plan (BCP) is a crucial component of project management in the oil and gas sector. It outlines strategies and procedures to ensure the continuity of critical operations in the event of unforeseen disruptions, such as natural disasters, equipment failures, or geopolitical events.

Key Elements of a Business Continuity Plan in the Oil and Gas Sector:

1. Risk Assessment: A risk assessment identifies potential risks and vulnerabilities that could impact the project's operations. This includes evaluating external factors like weather events, political instability, and market fluctuations, as well as internal factors such as equipment failure or supply chain disruptions.

2. Critical Function Identification: The BCP identifies the critical functions and processes that are essential for the project's success. This includes oil exploration, drilling, production, refining, transportation, and distribution. By prioritizing critical functions, the plan ensures that resources and efforts are focused on maintaining them during a disruptive event.

3. Contingency Measures: The BCP establishes contingency measures to address potential disruptions. These measures may include backup power systems, redundant equipment, emergency response plans, and alternative supply chain arrangements. Contingency measures help mitigate risks and ensure the continuity of operations in the face of unexpected events.

4. Communication and Stakeholder Engagement: Effective communication is vital during a crisis. The BCP outlines communication protocols for internal teams, stakeholders, and external partners. It ensures timely and accurate dissemination of information to manage the situation collaboratively, maintain stakeholder confidence, and minimize the impact of the disruption.

5. Recovery Strategies: The BCP includes recovery strategies to restore operations to normalcy after a disruptive event. These strategies outline detailed steps to resume production, repair or replace damaged equipment, and restore normal supply chain operations. Adequate training, technology, and resource allocation are essential components of recovery strategies.

6. Testing and Training: Regular testing and training exercises are critical to ensure the effectiveness of the BCP. Simulated drills and scenario-based training sessions allow teams to practice their roles and identify areas for improvement. Testing also helps fine-tune the plan based on real-time feedback, enhancing its ability to respond effectively during an actual crisis.

7. Continuous Improvement: The BCP is a living document that evolves with changing project dynamics and lessons learned

from past experiences. Regular reviews and updates ensure that the plan remains aligned with current risk profiles, emerging threats, and industry best practices. Continuous improvement ensures that the BCP is a robust tool for minimizing disruptions and enhancing project resilience.

Business continuity planning in the oil and gas sector project management is essential to protect investment, maintain stakeholder confidence, and safeguard the long-term success of projects. By proactively identifying risks, implementing contingency measures, and prioritizing critical functions, the BCP ensures that the project can quickly adapt and recover from unforeseen events, minimizing downtime and optimizing operations.

Bidding and proposal (Estimation)

The Bidding and Proposal team plays a crucial role in the EPC (Engineering, Procurement, and Construction) sector of the oil and gas industry. This team is responsible for developing and submitting competitive bids and proposals to secure contracts for EPC projects. Their primary objective is to showcase the company's capabilities, technical expertise, and value proposition to potential clients. Here is a brief overview of the Bidding and Proposal team's role in the oil and gas sector:

1. Opportunity Identification: The Bidding and Proposal team monitors the market, industry trends, and project opportunities within the oil and gas sector. They proactively identify potential projects that align with the company's capabilities, strategic objectives, and target markets.

2. Pre-Qualification and Preparing Pre-Qualification Questionnaires (PQQs): Depending on the project requirements, clients may request pre-qualification of contractors before initiating the bidding process. The Bidding and Proposal team is responsible for completing and submitting PQQs, demonstrating the company's

qualifications, experience, financial stability, and adherence to health, safety, and environmental standards.

3. Proposal Development: When a project opportunity arises, the Bidding and Proposal team leads the development of detailed proposals. This involves assessing project specifications, understanding client requirements, and formulating a comprehensive response. The team coordinates with various departments, including engineering, procurement, construction, and cost estimation, to gather the necessary information and create a compelling proposal.

4. Cost Estimation and Pricing: The Bidding and Proposal team collaborates with cost estimators and finance departments to develop accurate cost estimates for the proposed project. They analyze project requirements, consider various cost factors, assess risks, and determine a competitive pricing strategy that aligns with the company's profit margin goals.

5. Technical and Commercial Documentation: The team prepares and compiles all the necessary technical and commercial documentation required for the proposal. This includes project methodologies, execution strategies, project schedules, resource plans, quality assurance plans, health and safety plans, and any other documents requested by the client.

6. Strategy and Value Proposition: The Bidding and Proposal team formulates a clear strategy and value proposition for the proposed project. They highlight the company's unique selling points, competitive advantages, and innovations that set it apart from other potential bidders. The team emphasizes how their expertise and capabilities align with the client's project goals and objectives.

7. Proposal Submissions and Negotiations: The Bidding and Proposal team ensures the timely submission of the proposal, adhering to the client's submission requirements and deadlines. They may engage in negotiations with clients to address queries,

provide clarifications, and further refine the proposal based on client feedback.

The Bidding and Proposal team plays a critical role in securing EPC contracts in the oil and gas sector. Their efforts are instrumental in building relationships with clients, demonstrating technical capabilities, and showcasing the company's commitment to delivering successful projects. By consistently crafting high-quality proposals, the team enhances the company's chances of winning contracts and contributing to the growth and success of the organization in the oil and gas industry.

Training

Technical Training:

Technical training is essential for professionals in various industries, especially those that require specialized knowledge, skills, and expertise. Such trainings focus on enhancing individuals' technical competencies and proficiency in specific tools, technologies, processes, or methodologies related to their field.

The importance of technical training can be summarized as follows:

1. Expertise Development: Technical training helps individuals develop a deep understanding and mastery of technical concepts, techniques, and best practices. This expertise is vital for performing complex tasks, problem-solving, and staying updated with evolving industry trends and advancements.

2. Improved Efficiency and Productivity: With proper technical training, professionals can optimize their workflows, apply efficient methods, and leverage tools effectively. This leads to improved productivity, faster task execution, and better resource utilization.

3. Quality and Accuracy: Technical training enables individuals to enhance the quality and accuracy of their work. They gain the

knowledge and skills required to perform tasks with precision, adhere to industry standards, and deliver outputs that meet or exceed expectations.

4. Adaptability and Flexibility: Technical training provides professionals with the agility to adapt to changing technologies and work environments. They stay abreast of emerging tools and techniques, enabling them to embrace new challenges and opportunities while staying competitive in their field.

Soft Skills Training:

Soft skills training focuses on developing interpersonal, communication, and behavioral skills that are non-technical but crucial for professional success. These skills, often known as "people skills" or "personal skills," are valuable in any work environment and across various job roles.

SOFT SKILLS

However technically skilled a person can be, without honing the soft skills, it is very difficult to plan a career path.

many organizations provide soft skill training, but it is up to the individual to sharpen these skills and exhibit them.

Soft skills refer to a set of personal attributes and interpersonal skills that enable individuals to interact effectively and harmoniously with others in a professional setting. These skills are often intangible, as they are not directly related to technical expertise, but they are essential for success in various professional roles. Here are some important soft skills and their elaboration:

1. Communication Skills:

Effective communication involves expressing ideas clearly and listening actively. It encompasses verbal, written, and non-verbal communication. Strong communication skills facilitate effective collaboration, conflict resolution, and the ability to convey complex ideas simply and understandably.

2. Leadership Skills:

Strong leadership skills involve the ability to inspire, motivate, and guide others toward a common goal. This includes providing clear direction, delegating tasks effectively, empowering team members, and fostering a positive and inclusive work environment. Effective leaders can influence and inspire others to achieve collective success.

3. Teamwork and Collaboration:

Teamwork and collaboration skills enable individuals to work effectively with others, contribute their strengths, and cooperate toward a shared objective. This includes active participation, open-mindedness, willingness to compromise, and effective communication within a team.

4. Adaptability and Flexibility:

In today's rapidly changing work environments, adaptability and flexibility are crucial. These skills involve the ability to embrace change, adjust to new situations, and remain open to learning and growth. Adaptable individuals can quickly respond to evolving circumstances and navigate challenges with resilience.

5. Problem Solving and Critical Thinking:

Proficient problem-solving skills involve the ability to analyze complex situations, identify patterns, and develop effective strategies for resolution. Critical thinking skills enable individuals to evaluate information objectively, consider multiple perspectives, and make informed decisions based on sound reasoning.

6. Time Management and Organization:

Time management skills enable individuals to prioritize tasks, meet deadlines, and maximize productivity. Effective organizational skills involve maintaining a structured approach to work, managing resources efficiently, and keeping track of important details.

7. Emotional Intelligence:

Emotional intelligence encompasses the ability to understand and manage one's emotions, as well as empathize with others. It involves self-awareness, self-regulation, social awareness, and relationship management. Strong emotional intelligence enhances interpersonal relationships, conflict resolution, and overall teamwork.

8. Presentation and Public Speaking:

The ability to deliver clear and engaging presentations, as well as speak confidently in public, is a valuable skill. Effective presentation skills involve logically organizing information, using visual aids effectively, and engaging the audience through effective verbal and non-verbal communication.

9. Creativity and Innovation:

Creative thinking skills involve generating and implementing fresh and original ideas. Innovation skills enable individuals to identify opportunities for improvement, think outside the box, and propose innovative solutions that drive progress and transformation.

10. Networking and Relationship Building:

Networking skills involve building and maintaining positive professional relationships. This includes effective communication, active listening, and the ability to connect with others. Strong networking skills facilitate collaboration, information sharing, and potential career growth opportunities.

Developing and honing these soft skills can enhance professional growth, improve interpersonal relationships, and contribute to overall success in various professional endeavors.

Technical writing

Most of the engineers I have encountered are not strong in technical writing. Only those who are good at this grow on the ladder.

I strongly recommend every engineer to develop this skill as well as good communication skills.

Essentials of Technical Writing:

Technical writing is a specialized form of writing that focuses on conveying complex technical information clearly and understandably. Here are the essentials of technical writing:

1. Audience Analysis:

   - Understand the target audience and their knowledge level, expertise, and familiarity with technical concepts.
   - Consider their needs, goals, and potential challenges in comprehending the subject matter.
   - Adapt the writing style, vocabulary, and level of detail to best suit the intended audience.

2. Clarity and Simplicity:

   - Use clear and concise language to communicate technical information effectively.
   - Avoid jargon, acronyms, or technical terms that may confuse or alienate the audience.
   - Break down complex concepts into bite-sized information using plain language.
   - Structure the content logically, using headings, subheadings, and bullet points for clarity and easy navigation.

3. Visual Communication:

   - Utilize visuals like diagrams, charts, graphs, or illustrations to enhance understanding.

- Ensure visuals are accurate, labeled appropriately, and support the written content.
- Use consistent formatting and layout to create visually appealing documents.
- Provide clear captions or explanations for visuals to aid comprehension.

4. Organization and Structure:

- Follow a logical and consistent structure in organizing information.
- Start with an introduction to provide context and purpose.
- Use headings and subheadings to break down the content into sections for easy navigation.
- Present information in a sequential and orderly manner.
- Summarize key points and provide a conclusion or summary at the end.

5. Plain Language and Active Voice:

- Use simple language that is easy to understand.
- Write in an active voice, which makes the writing more engaging and direct.
- Keep sentences short and to the point.
- Define technical terms or provide explanations when necessary.

6. Accuracy and Precision:

- Ensure the accuracy and validity of the technical information presented.
- Verify facts, figures, and data using reliable sources.
- Use precise language to avoid ambiguity or misinterpretation.
- Provide citations or references for external sources used.

7. User-Centric Approach:

- Focus on the user's perspective and address their needs and concerns.

- Anticipate potential questions or areas of confusion.
- Provide clear and detailed instructions or procedures for using products or systems.
- Include troubleshooting tips or FAQs to help users overcome common issues.

8. Revision and Editing:

- Revise and edit the content for clarity, coherence, and grammatical correctness.
- Proofread for spelling, punctuation, and grammar errors.
- Eliminate unnecessary or redundant information.
- Ensure consistency in terminology and formatting.

9. Documentation Maintenance and Updates:

- Keep technical documents up to date with the latest information and changes.
- Regularly review and revise documents as needed.
- Establish version control and change management processes.
- Archive outdated or obsolete documentation.

10. Collaboration and Feedback:

- Collaborate with subject matter experts (SMEs) to ensure accuracy and completeness.
- Seek feedback from users or stakeholders to improve the documentation.
- Incorporate user feedback and suggestions to enhance usability.

By following these essentials of technical writing, you can create clear, concise, and user-friendly technical documents that effectively communicate complex information to your target audience.

Business Development

Business development in the EPC (Engineering, Procurement, and Construction) market within the oil and gas sector involves strategies and activities aimed at identifying new opportunities, fostering client relationships, and securing contracts for project execution. Here's an overview of how business development typically occurs in this industry:

Market Research and Analysis:

Business development teams conduct extensive research to identify potential clients, understand market trends, and assess industry needs. They analyze market dynamics, competitor positioning, and upcoming projects to identify opportunities for growth and expansion.

Building Relationships:

Establishing strong relationships with clients, industry stakeholders, and strategic partners is fundamental in the EPC market. Business development professionals actively network and engage in industry events, conferences, and forums to connect with potential clients and develop valuable partnerships.

Lead Generation and Proposal Development:

Lead generation is a core aspect of business development. Teams identify and qualify potential projects, assess their feasibility, and develop tailored proposals that highlight the company's expertise, capabilities, and value proposition. This involves collaborating with technical and engineering teams to address client requirements and develop innovative solutions.

Negotiation and Contracting:

Once a potential project opportunity is identified, business development professionals engage in negotiations with clients to finalize contract terms, scope of work, and pricing. This involves navigating commercial discussions, ensuring regulatory compliance, and addressing any contractual concerns to reach mutually beneficial agreements.

Strategic Alliances and Joint Ventures:

In the EPC market, companies often enter into strategic alliances and joint ventures to combine their strengths and resources, expanding their reach and capabilities. Business development teams actively explore partnership opportunities and engage in discussions to establish collaborative ventures that maximize project delivery and growth potential.

Market Expansion and Diversification:

Business development efforts also focus on expanding into new geographical markets and diversifying service offerings. This may involve exploring emerging markets, identifying niche sectors, and adapting service offerings to cater to evolving client needs and market demands.

Relationship Management and Client Satisfaction:

Even after securing a project, business development teams play a crucial role in maintaining healthy client relationships. They ensure client satisfaction, address concerns, and identify opportunities for continued collaboration or repeat business. This proactive approach helps foster long-term partnerships and generates referrals for future projects.

Continuous Improvement and Adaptation:

Business development professionals closely monitor market dynamics, industry trends, and client feedback to continually improve strategies and adapt to changing market conditions. They seek innovative approaches, embrace new technologies, and acquire insights from project experience to refine business development methodologies and maximize success.

In summary, business development in the EPC market within the oil and gas sector requires a proactive and strategic approach. It involves extensive market research, relationship-building, lead generation, proposal development, negotiations, and ongoing relationship management. By continuously adapting and pursuing new opportunities, businesses can sustain growth, secure projects, and establish their position in this competitive industry.

CHAPTER 10

Latest Trends in the EPC Project Management Domain

AGILE

AGILE: Empowering Success through Flexibility and Collaboration

AGILE has become a transformative approach in the world of project management. With its emphasis on adaptability, collaboration, and customer-centricity, AGILE methodologies have revolutionized how teams work together to achieve success.

At its core, AGILE is all about flexibility. Traditional project management methods often follow a rigid plan from start to finish, leaving little room for adjustments along the way. AGILE, on the other hand, embraces change and acknowledges that requirements evolve throughout a project's lifespan. This mindset allows teams to respond swiftly to new insights, unexpected challenges, and emerging opportunities.

Collaboration is another key aspect of AGILE. Cross-functional teams work closely together, fostering constant communication, knowledge sharing, and collaboration. Through daily stand-up meetings, regular feedback loops, and ongoing transparency, AGILE empowers teams to align their efforts, address issues promptly, and work towards a common goal. It cultivates an environment where everyone's expertise is valued, and decisions are made collectively.

AGILE also places a strong emphasis on customer satisfaction. By involving customers throughout the development process, AGILE methodologies ensure that their needs and expectations are met. This iterative approach allows for continuous feedback, resulting in products or services that truly align with customer desires. The focus is on delivering value early and often, taking into account customer feedback to refine and improve the outcome.

The benefits of embracing AGILE methodologies are numerous. Teams experience increased productivity, improved efficiency, and reduced risks. By breaking projects into smaller, manageable deliverables, AGILE promotes a sense of accomplishment and motivation. Failures or setbacks are seen as opportunities for learning, leading to continuous improvement in processes and outcomes. AGILE empowers teams to adapt to changing circumstances quickly, pivot when needed, and successfully navigate uncertain and dynamic environments.

Adopting AGILE requires a shift in mindset and a commitment to collaboration and flexibility. It necessitates an organizational culture that values communication, trust, and continuous learning. While AGILE may not be the right fit for every project or organization, it has proven to be a game-changer for those willing to embrace its principles.

In today's fast-paced and unpredictable business landscape, AGILE methodologies provide a roadmap to success. By focusing on flexibility, collaboration, and customer-centricity, AGILE enables teams to deliver impactful results and stay ahead of the curve. It's an empowering approach that empowers individuals and organizations to adapt, thrive, and achieve their goals in an ever-changing world.

SCRUM

SCRUM: Accelerating Success through Collaboration and Iterative Excellence

SCRUM has emerged as a powerhouse methodology in the realm of project management. With its emphasis on collaboration, adaptability, and iterative excellence, SCRUM revolutionizes the way teams work together to achieve outstanding results.

At its core, SCRUM is all about teamwork and collaboration. In a traditional project management setting, teams often follow a linear approach, where tasks are delegated, and progress is measured at milestones. However, SCRUM takes a different path. It brings together cross-functional teams, promotes daily communication, and fosters a culture of collective ownership and shared responsibility.

The heart of SCRUM lies in its iterative approach. Projects are divided into short, time-boxed iterations, called sprints. This enables teams to deliver tangible results rapidly and gather feedback early on, allowing for continuous improvement. With each sprint, teams adapt, refine, and build upon their learnings, providing added value with each iteration.

In SCRUM, roles are well-defined, with a clear distinction between the Product Owner, Scrum Master, and Development Team. The Product Owner ensures the project's vision is realized, shaping the backlog and prioritizing deliverables based on customer needs. The Scrum Master facilitates the SCRUM process, removing obstacles and fostering a productive environment. The Development Team collaboratively executes the work, utilizing their skills and expertise to bring the project to fruition.

SCRUM places significant importance on transparency and communication. Daily stand-up meetings ensure that team members are aligned, comfortable sharing progress, and aware of any impediments. The SCRUM board, with its task cards and visual representations, offers a clear snapshot of project status, promoting transparency and fostering a collective understanding of progress.

The benefits of SCRUM are substantial. Teams experience increased productivity, improved project predictability, and reduced risks. By embracing iterative development, SCRUM allows for the early detection of issues and facilitates timely course corrections. The incremental approach enables teams to deliver tangible value, keeping stakeholders engaged and satisfied.

SCRUM demands a shift in mindset and cultural adaptation. It thrives in environments where collaboration, open communication, and continuous learning are encouraged. SCRUM is not a one-size-fits-all solution, but rather a dynamic methodology that can be tailored to fit the unique needs and requirements of each project and organization.

In today's rapidly evolving business landscape, SCRUM empowers teams to navigate uncertainty with confidence, ensuring they stay on track and deliver exceptional results. By fostering collaboration, iteratively shaping projects, and embracing adaptability, SCRUM sets the foundation for success and drives projects toward excellence.

With SCRUM, the possibilities are endless. It's a proven framework that helps organizations adapt, thrive, and create extraordinary outcomes in a complex and ever-changing world.

DIFFERENCE BETWEEN AGILE AND SCRUM

AGILE and SCRUM are related, but they are not the same. AGILE is a broad project management philosophy or approach, while SCRUM is a specific framework within the AGILE methodology. Here are the key differences between AGILE and SCRUM:

1. Scope and Extensiveness:

- AGILE: AGILE is a comprehensive project management philosophy that encompasses various methodologies and frameworks. It focuses on adaptability, collaboration, and customer-centricity.
- SCRUM: SCRUM is a specific framework within the AGILE methodology. It provides a structured approach for managing complex projects, particularly those with rapidly changing requirements.

2. Structure and Framework:

- AGILE: AGILE does not have a strict structure or set of rules. Instead, it emphasizes core principles such as flexibility, continuous improvement, and customer collaboration.
- SCRUM: SCRUM provides a well-defined framework with specific roles, events, artifacts, and guidelines. It breaks projects into time-boxed iterations known as sprints, with clearly defined roles like the Product Owner, Scrum Master, and Development Team.

3. Focus and Application:

- AGILE: AGILE can be applied to a wide range of projects and industries, including software development, manufacturing, marketing, and more. It is adaptable and allows teams to tailor their principles and methodologies to fit their specific needs.
- SCRUM: SCRUM is primarily used in software development projects but can also be applied to other areas. It focuses on iterative development, collaboration, and delivering incremental value to the customer.

4. Team and Roles:

- AGILE: AGILE emphasizes cross-functional teams and collaboration. Team members often work together to achieve project goals, with an emphasis on self-organization and shared responsibility.
- SCRUM: SCRUM defines specific roles within the framework. The Product Owner is responsible for defining and prioritizing the product backlog, while the Scrum Master facilitates the SCRUM process. The Development Team is responsible for executing the work.

5. Documentation and Artifacts:

- AGILE: AGILE promotes lightweight documentation and values working software or tangible results over comprehensive documentation.
- SCRUM: SCRUM uses specific artifacts such as the product backlog, sprint backlog, and burndown charts to track progress, plan iterations, and communicate project status.

In summary, AGILE is a broader project management philosophy, while SCRUM is a specific framework within the AGILE methodology. SCRUM provides a structured, time-boxed approach to project management, emphasizing iterative development and collaboration. AGILE, on the other hand, encompasses a wider range of methodologies and approaches, allowing teams to tailor their project management practices to their specific needs.

Digital twin

Digital Twin: Revolutionizing the EPC Industry for Efficient Project Execution

The EPC (Engineering, Procurement, and Construction) industry is no stranger to complexities and challenges, from large-scale infrastructure projects to complex plant constructions. In this dynamic landscape, the integration of digital twin technology is transforming the way EPC projects are executed, ensuring efficiency, accuracy, and improved project outcomes.

Digital twin technology provides an invaluable platform for EPC companies to virtually represent and simulate the entire project lifecycle. By creating a digital replica of the physical assets and systems involved, EPC professionals can monitor, analyze, and optimize project performance in real time, gaining valuable insights and predictions.

Throughout the design and engineering phases, the digital twin enables collaborative engineering, facilitating multidisciplinary collaboration and concurrent design processes. Various teams can work together seamlessly, ensuring coordination, reducing errors, and enhancing project quality. By simulating different scenarios, design modifications can be tested virtually, identifying potential issues and optimizing engineering solutions, leading to cost and time savings.

During procurement and supply chain management, the digital twin fosters transparency, streamlining the process by visualizing and optimizing the flow of materials and equipment. EPC professionals can monitor inventory, track deliveries, and perform real-time quality checks, ensuring timely availability of resources and minimizing delays. Integration with supplier networks and advanced analytics empowers proactive decision-making, reducing procurement risks and enhancing supply chain efficiency.

As EPC projects move into the construction phase, the digital twin facilitates enhanced project execution. Real-time monitoring of key project parameters, such as progress, safety, and quality, ensures greater control and early detection of potential issues. By integrating Internet of Things (IoT) sensors and data, the digital twin captures project performance metrics, enabling predictive analytics and proactive maintenance. Construction progress can be visualized, analyzed, and optimized, optimizing resource allocation and minimizing rework.

Post-construction, the digital twin continues to provide value by enabling operations and maintenance optimization. By connecting the physical infrastructure with its digital counterpart, operations teams can monitor performance, detect anomalies, and evaluate asset health in real time. Predictive analytics and machine learning algorithms empower early fault detection, reducing downtime and improving asset reliability. The digital twin serves as a central repository of asset documentation and historical data, enhancing maintenance planning and decision-making for future projects.

The adoption of digital twin technology in the EPC industry comes with its challenges, including data management, system integration, and cybersecurity considerations. However, the benefits are enormous. EPC companies can drive efficiency, improve risk management, and deliver higher-quality projects with reduced cost and schedule overruns.

By embracing digital twin technology, the EPC industry can redefine project execution, optimize resources, and enhance collaboration. It paves the way for a more data-driven and intelligent approach, revolutionizing how projects are conceived, designed, constructed, and managed. The digital twin empowers EPC professionals to unlock unprecedented efficiency and deliver the infrastructure and facilities of the future.

Influence of AI in EPC sector

The Influence of Artificial Intelligence (AI) on the Future of the EPC Industry

Artificial Intelligence (AI) is poised to revolutionize the EPC (Engineering, Procurement, and Construction) industry, fundamentally transforming the way projects are planned, executed, and managed. As AI technologies continue to advance, their influence in the EPC industry will be profound, bringing numerous benefits and creating new opportunities for innovation. Here are some of the keyway's AI that will shape the future of the EPC industry:

1. Design Optimization: AI algorithms can analyze vast amounts of data, enabling EPC professionals to identify design optimizations, improve efficiency, and reduce costs. AI-powered design tools can generate optimal layouts, suggest material choices, and automate repetitive design tasks, accelerating project delivery and enhancing design quality.

2. Predictive Analytics: AI algorithms can leverage historical data, real-time inputs, and sensor data to predict project outcomes, identify potential risks, and optimize decision-making. By analyzing patterns and trends, AI-powered predictive analytics can help forecast project timelines, resource requirements, and potential bottlenecks, enabling proactive mitigation strategies.

3. Intelligent Automation: AI technologies, such as robotic process automation (RPA) and cognitive automation, can automate repetitive tasks, freeing up time for EPC professionals to focus on more complex and value-added activities. Intelligent automation can streamline documentation management, contract administration, and data processing, minimizing errors and optimizing workflow efficiency.

4. Quality and Safety Enhancement: AI-powered computer vision systems can analyze visual and sensor data to detect potential safety hazards, monitor construction progress, and ensure compliance with safety standards. AI can also analyze real-time data from various sources to detect quality issues, reducing rework and enhancing overall project quality.

5. Supply Chain Optimization: AI algorithms can optimize the supply chain, ensuring timely delivery of materials and equipment and minimizing inventory costs. AI can analyze historical purchasing patterns, market trends, and supplier performance to optimize procurement decisions, negotiate contracts, and identify potential cost savings.

6. Intelligent Asset Management: AI can enable smarter asset maintenance strategies by analyzing sensor data, performance metrics, and historical maintenance records. Predictive maintenance powered by AI algorithms can optimize maintenance schedules, detect anomalies, and predict failures, reducing downtime and maximizing asset reliability.

7. Augmented Decision-making: AI-powered decision support systems can provide engineers and project managers with real-time insights, recommendations, and scenario analyses. By integrating powerful AI algorithms with project management frameworks, EPC professionals can make better-informed decisions, mitigate risks, and achieve optimal project outcomes.

While the adoption of AI in the EPC industry presents immense opportunities, it also poses challenges related to data privacy, ethics, and skills development. Organizations must address these challenges to unlock the full potential of AI in the EPC sector.

AI's influence in the EPC industry will be transformative. By leveraging AI technologies, EPC companies can optimize processes, enhance project outcomes, and drive innovation. Embracing AI in the EPC industry will empower professionals to work smarter, faster, and more efficiently, ultimately shaping the future of project delivery and asset management.

Reality Capture

What is Reality Capture?

Reality capture is not just one technology but a whole group of technologies revolutionizing the construction tech industry. It's all based around the premise that being able to scan a building or object, then transfer that scan into a modeling or visualization plan, can help better represent what is taking place on the job site.

There are several ways that reality capture software is making a significant impact on the industry, but even with that cursory explanation, it can appear obvious why it might be so important.

The idea that you can 3D scan something on a construction site, gather all the data points, and then be able to create 3D models that anyone all over the globe can access is astounding. It has allowed for real-time analysis, and the ability for all stakeholders to get together in a virtual room and go through the current state of the project–even if the architect is in New York City, the engineer in Shanghai, and the owner in San Francisco.

Drone Technology

Drone technology is becoming ubiquitous on the job site, especially on large sites spanning large areas, such as large commercial projects, industrial projects, and infrastructure projects. As drones become more and more affordable and easier to operate–and more intelligent and better at capturing reality–they can be an almost constant site buzzing over construction sites.

These drones create a map of the site, using a coordinate system, and can capture data in a point cloud. From the information that these drones generate, entire maps of job sites are manufactured and transferred through the reality capture tool into a BIM product. In BIM (Building Information Modeling) a wireframe is constructed digitally for the project.

BIM modeling

From that BIM modeling, all stakeholders in a project, whether they're down the street or across the country, can see in real time the exact status of the project. Some BIM modeling can be as detailed as to show the individual studs and even electrical work. This all helps to save time and expense as the project continues because these stakeholders can identify problems early and address them while they're small.

Far beyond the time that the drones land and are shut down, architects and engineers can access the BIM data to see the progress of the site, locate areas where there may be potential problems, look for oversights and mistakes, and prevent disasters before they happen.

LiDAR Scanning

LiDAR, also known as laser scanning, fires pulses of laser light and then measures the amount of time it takes for the light to come back. You may have heard of LiDAR technology being used in archaeology fields to locate ruins in the thick undergrowth of Mexico or the Amazon. But it's also used on a construction site for the same effect, with its 3D laser scanning producing highly accurate data sets.

LiDAR can be used from the ground or the air, so it is seen in surveying equipment on the ground, in drones in the air, or even in low-flying aircraft–especially if using the LiDAR to survey a large span like a highway or bridge under construction.

A step ahead of its predecessor, photogrammetry software (a type of 3D scanning that uses photography and triangulation to create 3D models), LiDAR is pushing the boundaries of what can be accomplished with the technology that we have right now. And the bright news is that this is only the beginning, as newer technologies based on LiDAR are being tested even now.

LIDAR is being used in construction in many ways. Its penetrating lasers can help to map out a job site before the land-clearing process has gone underway, to identify obstacles and even structures that exist below the foliage. It can even be used on clear-cut terrain to locate underlying structures such as culverts, water mains, and the foundations of buildings that have been covered in earth. All of this goes to make the process of preparing a job site for work so much easier.

On-the-Ground Robots

We live in an age where robots move freely on job sites. While not as common sights as drones, robots, often tracked vehicles about the size of a large cooler, will roam the job site in a way similar to a Roomba. These roaming robots are recording more than just what can be viewed from aerial cameras and laser scanning, they are recording indoor things like positions of fixtures, columns, and even electrical outlets.

It's safe to say that any large construction project shortly is going to have these roaming robots traversing the site and making highly detailed maps of the interiors of buildings.

These robots, as they create detailed maps of structures, are building a real-time map from the ground level of the job site. This is different from the work that can be accomplished with drones, as the on-the-ground robots can work in buildings that already have roofs built. The mapping that these robots do can be uploaded and turned into a BIM graphical interface.

With these four methods of reality capture, one can continue to leverage point clouds and work with reality capture technology now and in the future.

The Gas Industry and Data Visualization Tools

The gas industry is a vital sector that plays a critical role in powering homes, industries, and vehicles across the globe. The industry is constantly evolving, with new technologies and advancements emerging regularly. To keep up with these changes and make informed decisions, it is essential to leverage data and visualization tools. In this article, we will explore how gas industry professionals can use visualization tools to gain valuable insights and make data-driven decisions.

Why Visualization Tools Matter in the Gas Industry

In the gas industry, huge amounts of imported data are generated every day from a wide range of varied and disparate data sources. From production and consumption to safety and compliance, data is at the core of every aspect of the industry. However, it is challenging to make sense of this data without effective tools for data visualization.

Visualization tools can help gas industry professionals understand complex data sets, identify trends, and gain insights quickly. Data analytics visualization tools help to transform data into meaningful and easily understandable visualizations; these analytics tools include features that make it easier to spot patterns and make informed decisions.

Visualizations also help to improve communication within the industry. They make it easier to share information and insights with stakeholders, including management, customers, and regulators-all of which are part of business intelligence leading to actionable insights.

Types of Visualization Tools in the Gas Industry

There are several types of data visualization software that gas industry professionals can use to gain insights and make data-driven decisions. Here are some of the most common ones:

- Geographic Information Systems (GIS): GIS tools enable the visualization of data on maps. This includes information on gas pipelines, storage facilities, production sites, and other gas-related infrastructure. GIS tools help to identify potential issues or bottlenecks in the gas supply chain and make informed decisions.
- Dashboards: Dashboards provide an overview of key performance indicators (KPIs) and other relevant data in real-time. This includes data on gas production, consumption, and reserves, as well as safety and compliance metrics. Dashboards allow gas industry professionals to monitor performance and identify trends quickly, creating visual interpretations of data.
- 3D Visualization: 3D visualization tools help gas industry professionals visualize gas infrastructure, including pipelines and storage tanks, in three dimensions. This allows for better planning and analysis of gas infrastructure projects.

- Machine Learning (ML) and Artificial Intelligence (AI) Tools: ML and AI tools can help identify patterns in large data sets and make predictions about future trends. These tools can be used to identify potential issues before they occur and optimize gas production and distribution

Benefits of Visualization Tools in the Gas Industry

The benefits of using visualization tools in the gas industry are numerous. Some of the key advantages include:

- Improved Decision-Making: Visualization tools provide gas industry professionals with valuable insights that can inform decision-making. This includes identifying trends, predicting future outcomes, and highlighting potential issues.
- Enhanced Safety: Visualization tools can help gas industry professionals identify safety risks and take corrective action before accidents occur. This includes monitoring gas pipelines, storage facilities, and other infrastructure for potential issues.
- Increased Efficiency: Visualization tools allow industry professionals to optimize gas production and distribution, reducing waste and improving efficiency. This includes identifying areas where production can be increased and reducing the time needed to identify and resolve issues.
- Improved Communication: Visualization tools make it easier to share information and insights with stakeholders. This includes sharing data with management, customers, and regulators, as well as collaborating with other professionals in the industry.

The gas industry is constantly evolving, and visualization tools are an essential component of staying ahead of the curve. By leveraging these tools, gas industry professionals can gain valuable insights, improve decision-making, and increase safety and efficiency. With the right visualization tools, gas industry professionals can make informed decisions that drive success and help to power the world.

IoT (internet of things)

IoT has many ways of implementation in the oil and gas industry, including collecting, transferring, and analyzing raw data in real-time. This helps companies to have more valuable information about the different processes.

In the oil and gas industry, the Internet of Things can be used to manage operations and optimize processes. One advantage is predictive maintenance. By using sensors to monitor equipment such as drilling machinery, pipelines, and storage tanks in real-time, companies can detect potential issues before they lead to costly breakdowns. This proactive approach will reduce downtime and save substantial maintenance costs.

Furthermore, IoT solutions enhance safety by enabling remote monitoring of hazardous environments, ensuring that workers are not exposed to unnecessary risks. In summary, IoT's transformative potential in the oil and gas sector extends far beyond data management, making operations more efficient, cost-effective, and safer.

As a tool, the IoT in the oil and gas industry brings many benefits to the field and empowers professionals and business owners to make progress by using these advantages.

But how exactly do the Internet of Things and IoT solutions help the industry, and what to expect when using it?

The Importance of IoT in the Oil and Gas Sector

Of course, innovative technologies play a hugely important role in modern businesses, but this is not the only way they affect professional processes. Technologies, and solutions that are based on these technologies, are a way to achieve more, have better results, and optimize certain processes. But how?

In this dynamic industry where remote and harsh environments are the norm, IoT-enabled sensors provide real-time data on equipment performance, environmental conditions, and safety parameters.

This data enables companies, known for exceptional expertise in software development, to develop advanced monitoring systems. These systems help in optimizing production processes by monitoring reservoirs and pipelines for maximum resource extraction while ensuring the reliability of critical equipment.

Additionally, innovative solutions have been instrumental in reducing downtime and maintenance costs for businesses operating in challenging environments.

Advantages of IoT in the Oil and Gas Industry

The adoption of Internet of Things (IoT) technology in oil and gas has brought about a multitude of advantages, revolutionizing the way companies operate in this sector.

Sensors and connected devices are deployed throughout oil and gas facilities, allowing for real-time monitoring of equipment and processes. This data-driven approach enables companies to detect anomalies, predict equipment failures, and optimize maintenance schedules-- leading to reduced downtime and lower operational costs. Additionally, IoT facilitates remote monitoring and control--reducing the need for on-site personnel in hazardous environments and improving safety conditions for workers.

The Internet of Things can help to improve both safety and environmental sustainability within the oil and gas industry. By monitoring environmental conditions, such as air quality and water levels, companies can respond more rapidly to potential leaks or spills. This helps prevent accidents and minimizes the impact that operations have on the environment. Furthermore, integrating IoT into oil and gas operations promotes both operational excellence and an environmentally responsible approach.

Utilizing IoT Capabilities in Oil and Gas Enterprises

Leveraging the capabilities of the Internet of Things (IoT) has become increasingly integral to the operations of oil and gas enterprises. IoT technologies, such as sensors and connected devices, offer real-time data collection and monitoring capabilities across remote and critical infrastructure.

These innovations enable companies to enhance their asset management, optimize production processes, and improve safety measures. By harnessing IoT in oil and gas, enterprises can attain greater operational efficiency, reduce downtime, and make data-driven decisions, ultimately resulting in cost savings and increased competitiveness in this dynamic and resource-intensive industry.

The Role of IoT in Exploration and Production Processes

The oil and gas sector is rapidly adopting IoT technology to transform exploration and production processes. Forward-thinking companies are shifting their focus from using IoT devices like sensors to developing innovative strategies for leveraging the data these devices collect, creating smarter business models for greater success.

In seismic data acquisition and analysis, IoT-enabled sensors provide real-time insights into subsurface structures, accelerating exploration and improving reservoir modeling.

In drilling operations, IoT integration optimizes efficiency, reduces costs, and enhances safety through real-time monitoring and predictive maintenance. Overall, IoT is revolutionizing the oil and gas industry by streamlining processes and improving decision-making.

IoT in Action: Pipeline Monitoring and Maintenance

In the realm of pipeline monitoring and maintenance, IoT (Internet of Things) technology is reshaping industry practices. IoT devices are strategically placed along pipelines, continuously collecting data on parameters like pressure, temperature, and flow rates. This real-time data transmission to a central control system allows for careful monitoring and control of different essential aspects, as well as for rapid issue response, which will reduce the risk of downtime and environmental damage.

Additionally, machine learning algorithms enable predictive analytics, forecasting maintenance needs, and optimizing resource allocation, making pipeline management smarter, more efficient, and environmentally responsible.

Asset Integrity Management System (AIMS)

Asset Integrity Management System in the Oil and Gas Sector: Ensuring Safety, Reliability, and Efficiency of Critical Assets

In the oil and gas sector, asset integrity management is of paramount importance to ensure the safe, reliable, and efficient operation of critical assets throughout their lifecycle. An Asset Integrity Management System (AIMS) encompasses comprehensive strategies, practices, and processes that aim to maintain the integrity and performance of assets, ensuring compliance with regulations and industry standards. Here's a detailed exploration of the significance and benefits of an AIMS in the oil and gas sector:

1. Enhancing Safety and Risk Mitigation:

An AIMS establishes robust safety protocols and risk management practices to identify, assess, and mitigate risks associated with asset integrity. It helps prevent incidents, accidents, and operational disruptions by implementing proactive measures to identify potential hazards and address them before they escalate.

2. Ensuring Regulatory Compliance:

With stringent regulatory requirements in the oil and gas industry, an AIMS assists in maintaining compliance with safety standards, environmental regulations, and legal obligations. It ensures that assets adhere to prescribed guidelines, preventing costly penalties, regulatory consequences, and reputational risks.

3. Extending Asset Lifecycle and Optimal Performance:

Through preventive maintenance, condition monitoring, and continuous inspection practices, an AIMS helps extend the lifecycle of assets, maximizing their performance and operational efficiency. It facilitates timely identification of degradation, corrosion, or aging issues, enabling necessary interventions to eliminate or mitigate integrity risks.

4. Optimizing Maintenance and Inspection Activities:

An AIMS enables the development of optimized maintenance and inspection strategies, ensuring that resources are allocated appropriately based on asset criticality. It facilitates risk-based decision-making to enhance the effectiveness of inspection and maintenance practices, reducing operational costs while maintaining asset integrity.

5. Implementing Asset Monitoring and Data Analysis:

Utilizing advanced technologies and monitoring systems, an AIMS enables real-time or periodic asset monitoring, data collection, and analysis. This facilitates early detection of anomalies, deviations, or performance deterioration, allowing timely corrective actions and preventing unplanned downtime or failures.

6. Integration of Technology and Data Management:

An AIMS leverages digital technologies, such as the Internet of Things (IoT), sensors, and data analytics, to enhance asset monitoring, data management, and predictive maintenance. It fosters the integration of data from different sources, enabling effective decision-making and providing accurate asset performance insights.

7. Fostering Proactive Culture and Continuous Improvement:

An AIMS cultivates a proactive safety culture within organizations, encouraging employees to report potential issues, near misses, or deviations from standard procedures. It facilitates performance benchmarking and regular evaluation to drive continuous improvement in asset integrity management practices and processes.

In summary, an Asset Integrity Management System is vital in the oil and gas sector, providing a comprehensive framework to maintain the safety, reliability, and efficiency of critical assets. It enables proactive risk management, regulatory compliance, and optimized maintenance practices, leading to extended asset lifecycles, improved operational performance, and enhanced overall industry sustainability.

Latest trends in Quality Management

One of the latest trends in quality management is the adoption of technology and digital transformation. Organizations are leveraging advancements in technology to enhance and streamline their quality management processes. Some key developments in this area include:

1. Industry 4.0 and Internet of Things (IoT): IoT devices are increasingly being used to collect real-time data from equipment, processes, and products. This data can be analyzed to identify quality issues, optimize production, and enable timely corrective actions.

2. Artificial Intelligence (AI) and Machine Learning (ML): AI and ML algorithms are being employed to analyze vast amounts of quality-related data, identify patterns, and predict potential quality issues. This helps organizations proactively address problems before they occur and improve decision-making.

3. Big Data Analytics: With the exponential growth of data, organizations are using analytics tools to extract insights from diverse datasets. These insights enable better quality management decision-making, process optimization, and predictive analytics for quality control.

4. Cloud-Based Quality Management Systems (QMS): Cloud-based QMS platforms provide organizations with the ability to centralize quality-related data, streamline document control, automate workflows, and facilitate collaboration across different teams and locations.

5. Risk-Based Quality Management: Organizations are increasingly adopting a risk-based approach to quality management. This involves identifying potential risks to quality, prioritizing them based on their impact, and allocating resources effectively to mitigate those risks.

6. Supplier Quality Management: Emphasis is being placed on building strong relationships with suppliers and implementing robust supplier quality management processes. This includes supplier performance monitoring, transparency, and collaboration to ensure consistent quality throughout the supply chain.

7. Customer-Centric Quality: Organizations are recognizing the importance of customer satisfaction and incorporating it into their quality management strategies. Measures such as customer

feedback collection, sentiment analysis, and personalized quality experiences are gaining significance.

8. Continuous Improvement: Continuous improvement methodologies such as Lean Six Sigma, Total Quality Management (TQM), and Kaizen remain significant in quality management. Organizations are continuously seeking ways to eliminate waste, reduce defects, and improve overall quality.

Overall, the latest trends in quality management focus on leveraging technology, data-driven decision-making, risk management, and customer-centricity to drive improvement, enhance efficiency, and ensure superior product and service quality.

TITBITS

Email etiquettes

1. Use a Clear and Professional Email Address:
Ensure that your email address reflects professionalism, preferably using your name. Avoid using casual or unprofessional email addresses that may create a negative perception.

2. Write a Clear and Concise Subject Line:
Create subject lines that accurately represent the content of the email. A clear subject line helps recipients understand the purpose of your email and facilitates smooth communication.

3. Use Proper Greetings and Salutations:
Begin your email with a courteous greeting, such as "Dear Mr./ Ms." or "Hello [First Name],". Use appropriate salutations, such as "Sincerely" or "Best Regards," to end your email in a professional manner.

4. Be Mindful of Tone:
Emails can sometimes be misinterpreted, so choose your words carefully. Use a polite and professional tone and avoid using sarcasm or ambiguous statements that may cause confusion or offense.

5. Keep It Professional and Formal:
Maintain a professional tone throughout the email. Use complete sentences and proper grammar. Avoid the use of slang, abbreviations, or excessive emojis, as they are not suitable for professional communication.

6. Keep Your Email Brief and to the Point:
Respect the recipient's time by keeping your email concise and focused. Clearly state the purpose of the email and provide essential

information efficiently. Use bullet points or paragraphs to break up lengthy text, ensuring readability.

7. Use Correct Grammar and Proofread:

Always proofread your email for grammar and spelling errors before hitting the send button. A well-written and error-free email demonstrates attention to detail and professionalism.

8. Respond Promptly:

Strive to respond to emails in a timely manner. Even if you cannot provide a complete response immediately, acknowledge the receipt of the email and let the sender know when they can expect a detailed response.

9. Use Proper CC and BCC:

When addressing multiple recipients, use the CC (carbon copy) field for those who need to be informed and the BCC (blind carbon copy) field when recipients do not need to be visible to each other. Be mindful of sharing sensitive information.

10. Use Professional Signature:

Create an email signature that includes your full name, position, and contact details. This allows recipients to easily identify and contact you.

Remember, emails are a reflection of your professionalism and attention to detail. Following these email etiquette guidelines will help you effectively communicate and build positive relationships within your organization

CHAPTER 11

Project Closeout

The project closeout process marks the final phase of an EPC (Engineering, Procurement, and Construction) project in the oil and gas sector. It involves the systematic completion of project activities, ensuring that all deliverables have been met, and the project is ready for handover to the client. Here's a detailed write-up on the project closeout process:

1. Documentation Review:

- Conduct a thorough review of all project documentation to ensure completeness, accuracy, and compliance with regulatory requirements, industry codes, and client specifications.
- Verify that all engineering deliverables, including design drawings, calculations, specifications, and test reports, have been properly documented and archived.

2. Punch List Management:

- Develop and manage a punch list that captures any outstanding work items, deficiencies, or non-compliant items identified during inspection, testing, or commissioning.
- Collaborate with subcontractors and project teams to address and close out these punch list items promptly.

3. Final Inspections and Testing:

- Perform final inspections and testing to validate the quality, functionality, and compliance of all installed equipment, systems, and components.

- Ensure that all required documentation, including inspection reports, test certificates, as-built drawings, and operation manuals, are prepared, reviewed, and approved as part of the closeout process.

4. Commissioning and Start-up Activities:

- Coordinate any remaining commissioning activities required to ensure that all systems, equipment, and processes are successfully brought into operation.
- Conduct performance testing and functional checks to verify that the project meets the required performance criteria and operates as intended.

5. Contractual Obligations and Deliverables:

- Review the project contract and ensure that all deliverables, milestones, and contractual obligations have been met.
- Coordinate with project stakeholders, including the client, to complete any outstanding deliverables and resolve any outstanding contractual issues or claims.

6. Financial Closure:

- Finalize all financial aspects of the project, including the submission and review of final invoices, progress payment reconciliation, and resolution of any outstanding financial matters.
- Ensure that all financial records, including receipts, invoices, and financial statements, are properly documented and archived.

7. Lessons Learned and Documentation:

- Conduct a comprehensive "lessons learned" exercise to capture project insights, best practices, and areas for improvement. Document these lessons and share them with the project team for future projects.

- Compile final project documentation, including an organized project closeout report that encompasses all project activities, achievements, challenges, and key learnings.

8. Project Handover and Client Acceptance:

- Prepare all required handover documentation, including operation and maintenance manuals, warranties, training materials, and any other contractual deliverables.
- Facilitate the smooth transition of the project to the client, ensuring client acceptance and sign-off on project completion.

9. Project Evaluation and Review:

- Conduct a project evaluation with the project team and stakeholders to assess the project's overall performance against its objectives, timeline, budget, and quality.
- Identify areas of success and challenges, providing insights for future projects and continuous improvement.

The project closeout process in the EPC oil and gas sector is critical to ensure the successful completion of all project activities and the transfer of a fully operational and compliant project to the client. By effectively managing the closeout process, organizations can demonstrate their commitment to quality, client satisfaction, and industry standards.

"MEETINGS ARE WHERE HOURS ARE WASTED AND MINUTES ARE RECORDED"

Tips for conducting effective meetings and etiquette...

Prepare an agenda and circulate it well in advance to all the participants, so that they come prepared.

Appoint a moderator for the meeting, and the moderator carries the meeting as per the agenda. Allow every participant to express their points during the meeting.

Ideally, 1 hour should be the time limit for any meeting.

Avoid serving coffee or tea snacks during the meeting which districts the group.

Do not use a laptop during meetings, give attention only to the meeting.

One can relieve the participant, once their, portion of work is complete so that they can continue with their routine work

It is a good practice to have one day of the week as meeting meeting-free day.

Avoid planning meetings in the morning session (forenoon), because the employee will be at his peak of energy, in the morning. Plan meetings during the second half.

Do not deviate from the agenda as it will dilute the purpose of the meeting.

Ask someone to note down the minutes of the meeting.

The minutes of the meeting should mention the action date and the person to whom the action is assigned.

Request all participants to keep the mobile in silent mode.

It's a good practice to start the meeting with a value moment or a safety moment.

Do not get into arguments or debates during the meeting and ensure it's constructive.

Avoid long PPT presentations during the meeting unless otherwise it's required.

Introduce the participants if they are new to the organization.

Avoid business lunch during meetings and utilize lunch time, for interacting with the participants and developing rapport.

Ensure sufficient lighting, air conditioning, and water are available during the meeting.

For multi-center meetings work out a convenient time zone so that the participants are comfortable.

Dress code during the meeting is very important in particular when clients or vendors are participating in the meeting.

Finally, avoid planning too many meetings in a day and avoid planning meetings on Friday afternoon. Declare any one day in a week as a meeting-free day.

Conclusion

The EPC oil & gas sector is a challenging industry. One is required to update the skills and knowledge to climb up the ladder. Multi-skilling and multi-tasking are highly encouraged for success.

Remember, LIFE is not a SPRINT, it is a MARATHON